D0464998

MILITARY ROBOTS AND DRONES

Selected Titles in ABC-CLIO's
CONTEMPORARY
WORLD ISSUES
Series

For a complete list of titles in this series, please visit
www.abc-clio.com.

Books in the Contemporary World Issues series address vital issues in today's society, such as genetic engineering, pollution, and biodiversity. Written by professional writers, scholars, and nonacademic experts, these books are authoritative, clearly written, up-to-date, and objective. They provide a good starting point for research by high school and college students, scholars, and general readers as well as by legislators, businesspeople, activists, and others.

Each book, carefully organized and easy to use, contains an overview of the subject, a detailed chronology, biographical sketches, facts and data and/or documents and other primary source material, a directory of organizations and agencies, annotated lists of print and nonprint resources, and an index.

Readers of books in the Contemporary World Issues series will find the information they need to have a better understanding of the social, political, environmental, and economic issues facing the world today.

MILITARY ROBOTS AND DRONES

A Reference Handbook

Paul J. Springer

CONTEMPORARY WORLD ISSUES

 ABC-CLIO

Santa Barbara, California • Denver, Colorado • Oxford, England

Library of Congress Cataloging-in-Publication Data

Springer, Paul J.
  Military robots and drones : a reference handbook / Paul J. Springer.
    p. cm.
  Includes bibliographical references and index.
    ISBN 978–1–59884–732–1 (cloth : alk. paper) — ISBN 978–1–59884–733–8 (ebook) 1. Robotics—Military applications. 2. Military robots. 3. Drone aircraft. 4. Uninhabited combat aerial vehicles. I. Title.
UG450.S68   2013
623—dc23   2012031879

ISBN: 978–1–59884–732–1
EISBN: 978–1–59884–733–8

17 16 15 14 13   1 2 3 4 5

This book is also available on the World Wide Web as an eBook.
Visit www.abc-clio.com for details.

ABC-CLIO, LLC
130 Cremona Drive, P.O. Box 1911
Santa Barbara, California 93116-1911

This book is printed on acid-free paper (∞)

Manufactured in the United States of America

Contents

Preface

I vividly remember my introduction to the concept of military robots. The day after Christmas, 1985, my best friend Bobby Wright invited me to his home to see his newest toys, as was our time-honored holiday habit. He proudly displayed his new Optimus Prime Transformer robot, igniting a fascination in me that has remained to this day. While the notion of alien shape-shifting robots is certainly farfetched, military robots are rapidly changing their forms, functions, and applications on the battle-field. This work is intended as an examination of the state of the field of military robotics in 2012. It is heavily focused on the United States, in part because of the American leadership role in the development of the field, and in part because of the transparency of most U.S. military programs, particularly when they reach the deployment stage. While other nations are working to increase the capabilities of their military robots and drones, the United States is by far both the largest producer and user of military robots, and much of what the United States does will set the tone for at least the next few decades, if not longer.

Military robots offer tremendous potential utility, but their employment is also fraught with peril. While they potentially serve to keep one's own human forces away from danger, their existence also contains the potential to erode or destroy the fundamental laws and ethics of warfare that have governed interstate conflict since the seventeenth century. In part, this work is a cautionary tale, reminding the reader that revolutionary military technology has rarely been confined to the effects intended by its creators, and the field of robotics is no exception. While the Hollywood visions of self-aware robots rebelling against their creators is ludicrous at this time (and, incidentally, was the plot of the play that introduced the term "robot"), the notion that they will be used in new and terribly destructive fashions is all but a certainty. Only by attempting to actively manage the proliferation and utilization of these new devices will states have any chance of preventing the very worst possible outcomes.

Of course, no scholarly undertaking of this magnitude is ever undertaken alone, and this book is no exception. I must thank first and foremost my wife, Victoria, who has tolerated the late nights and bizarre robotic obsessions with good grace. I would also like to recognize my colleagues at Air University, Maxwell Air Force Base, for their support and input. Our faculty is a diverse mix of civilians and military personnel, bringing a broad variety of expertise into a single school. Thanks to all of their knowledge, I was able to pester experts in military theory and operators with years of combat experience at the same time. Of particular assistance in this manner were Everett Dolman, Mack Easter, Paul Hoffman, Ben Jacobson, Tom Kelly, James Kiras, Erik Kisker, J. T. LaSaine, Jeff Lin, Michael May, S. Michael Pavelec, John Reese, James Selkirk, Nathan Tarkowski, John Terino, and Cliff Theony. I owe each a substantial intellectual debt for their assistance and insights. I am also thoroughly indebted to ABC-CLIO's Pat Carlin, who not only offered me the opportunity to write this work but who also remained unflinchingly patient as my deadlines became all too flexible. Likewise, Robin Tutt's editorial expertise proved invaluable to the final product. Sethu Baskaran Natarajan and his team at PreMedia Global made major contributions to the finished work, and I appreciate their assistance. In particular, Sarah Wales-McGrath did fantastic work in preparing the final manuscript.

In my career, I have had tremendous opportunities to work with some of the best and brightest young military minds in the nation, both at West Point and at Air University. It is to all my students, past, present, and future, many of whom will have to confront the changes wrought by military robotics, to whom I humbly dedicate this volume. I thank them for their service and hope that I have made some small contribution to their success.

1

Background and History

The idea of a history of robotic warfare is in some ways a misnomer, as completely autonomous weapons have not yet been deployed on or over any battlefield. Many experts believe such a usage is only a matter of time, as several advanced militaries have already developed semiautonomous weapons, and dozens of nations are actively pursuing the acquisition of such machines. The United States currently leads in the technological innovation of autonomous weapons, but China, France, Germany, Great Britain, Israel, and Russia also have extremely advanced programs. The idea of substituting machines for humans on the modern battlefield simply has too much allure for any nation that expects to face substantial conflict in the future. In addition to preserving the lives of its citizens, the nation that adopts such technology may also obtain a substantial tactical advantage. Robots and drones also offer potential economic savings, as the production costs drop and the long-term usage costs tend to be far lower than manned systems.

Defining Robots and Drones

There are many definitions currently in use for the term "robot." Most variants consider a robot to be a machine capable of sensing its environment and reacting to it through independent decision-making capability (Perkowitz 2004, 4). This implies at least some interaction with the robot's surroundings, although it does not

1

expressly require the machine to be mobile or intelligent. It may be a stationary platform that reacts only in response to certain stimuli or under particular conditions. Simply stated, a robot does not need to fit the popular cultural expectation of a metallic humanoid performing mundane tasks on behalf of its creator. Robots come with virtually unlimited shapes, sizes, purposes, and costs. Those used by military forces have a dual purpose, in that their function may not be inherently military in nature, and many military robots are indeed off-the-shelf robots initially designed for other markets.

While a robot requires at least some decision-making abilities, a drone is a machine that performs a preprogrammed task with or without human interaction. A true drone does not make independent decisions, although it may appear to do so to the outside observer. Drones can perform their tasks without regard to their surroundings, as they have no inherent need for sensory input, and they also have no need for even the most rudimentary intelligence. In practical applications, many drones are outfitted with sensory gear, but it is not for the benefit of the machine. Rather, it is to enable the remote operation of the device by a distant human controller.

Many of the military systems in use today straddle the line between robots and drones, incorporating elements of both. These devices are typically referred to as "robotic," in that they have a limited degree of decision-making ability and interaction with their environment but are also under the command and control of a human operator. The concept of autonomy is a clear dividing point between drones and robots. To be truly autonomous, a machine must be allowed to perform its entire function without input from a human operator. Some of the most well-known systems incorporate limited autonomy, particularly unmanned aircraft. This is in part because the machine's reaction speed is greater than that of a human operator, particularly one attempting to fly an aircraft from a remote ground control station that receives limited sensory input from the vehicle. The most advanced systems can now take off and land completely independent of the actions of their operators, and a 2004 study demonstrated that a significant percentage of unmanned aerial vehicle (UAV) crashes were due to pilot error, suggesting that greater self-control might lead to less attrition in the inventory (Williams 2004). They can choose their own flight paths, deconflict airspace with fellow aircraft, both manned and unmanned, and select their own vectors to attack a target. They can track and follow other vehicles, whether in the air or on the ground. Perhaps most importantly, they can select appropriate onboard munitions, target them, and directly attack an enemy. Although the most advanced systems can perform all of these functions independently, the current use of unmanned systems retains a human in the decision-making function regarding the application of lethal force. However, the constantly improving nature of the machines involved suggests that in the near future, the systems may be granted the ability to decide for themselves whether to launch an attack.

The key enabling technology for robotic weaponry is computing power. For decades, computer scientists and programmers have sought to develop a machine that could emulate human cognition. Theoretically, it could perform the same tasks as a human mind, to include designing ever-more-advanced computers. Such a machine could be considered an artificial intelligence (AI) if its decision-making capability were as advanced as that of a human. While the traditional definition of AI requires that a human observer be unable to tell the difference between a computer's responses and those of a human, the reality is that computers may demonstrate AI despite making decisions in an entirely different fashion than their human operators. In particular, a computer is far more likely to behave in a rational, logical fashion, rather than an unpredictable, emotional state. For this reason, some theorists have argued that even the most advanced computers may not be able to engage in creative problem solving, as they will pursue only solutions that have been devised by their programmers. While a true AI has yet to be developed, many experts predict that it will be present by the year 2020 (Van Joolen 2000, 3). Once AI has been achieved, the speed of computer advances is likely to increase, as computers, unlike their human counterparts, can concentrate their entire capability on a single task and can continue to attack the same problem without facing the issues of fatigue or boredom.

Another important distinction within the field of robotic weapons comes from the terms "unmanned" and "remotely piloted." A vehicle is unmanned if it does not carry a human, but it can be considered remotely piloted only if an operator is actually controlling or guiding the movement of the vehicle. The current most advanced systems allow input from human operators, particularly regarding the release of ordnance, but do not require remote piloting during most operations. Some, like the RQ-4 Global Hawk, perform all flight operations independent of human guidance and accept input primarily in the form of mission orders. This approach has greatly reduced the number of pilot errors, though faulty programming has still caused some accidents. In comparison, many of the smaller unmanned platforms, regardless of the environment that they operate in, are completely remotely piloted, in the same manner as a radio-controlled toy car or airplane. These systems are usually much cheaper to produce and as a result are considered far more expendable.

There are some weapons systems that incorporate limited autonomy but that should not be considered robots or drones, even though they may include some robotic elements. While robots and drones are often considered expendable, in that their loss is not so costly as their corresponding manned platforms would be, they are not inherently designed to be single-use weapons. Their greatest utility comes from their ability to repeatedly perform missions considered too dangerous, dirty, or tedious for manned systems. Cruise missiles, for example, should not be considered robots or drones even though the most advanced cruise missiles can determine their own flight paths, including deliberate attempts to prevent detection or avoid enemy interception (Sundvall 2006, 7). Some can be given new targets while in-flight and rerouted subject to their limits of their range. Ultimately, cruise missiles are a single-use weapon and as such, do not qualify as robots or drones, regardless of their sophistication (Clark 2000, 3–4). Likewise, land mines, despite the rudimentary ability of many to detect and respond to external stimuli, are merely reflexive systems. The fact that many can distinguish between potential targets and explode only when touched by a metallic object, or when subjected to the weight of a vehicle, does not mean that they are robotic in nature. They are not autonomous weapons platforms at this time, although some designs have been created for land mines that would be able to move themselves and truly decide whether to attack a target. Such advanced systems may cross over into the robotic platform in the near future.

This chapter will provide an overview of the long history of humanity's attempts to develop autonomous weapons systems.

Although the actual fielding of such machines did not occur until the late 20th century, the antecedents of autonomous weapons shaped their eventual design and employment. These ancient devices provided inspiration for what might someday be possible and a vision of how it should be used. As such, it is valuable to examine the history of humanity's attempts to replace people in warfare. Also, because so much robotic technology is devoted to nonmilitary uses, it is helpful to examine the development of the robotic field over time.

Early Descriptions and Imagined Machines

Arthur C. Clarke once famously observed that "Any sufficiently advanced technology is indistinguishable from magic" (Clarke 1973, 21). To a population unfamiliar with the function or development of a high-technology device, magic is a perfectly acceptable explanation. This concept has merit when examining ancient conceptions of warfare, particularly with regard to the use of autonomous machines. Myths from around the globe include references to nonhuman armies, whether composed of reanimated corpses, artificial life forms, supernatural beings, or machines created by diabolical inventors. In the same way that advanced technology and magic become indistinguishable under certain circumstances, the same could be said of the real world and the supernatural to ancient observers. What modern civilization would consider superstition or imaginary, ancient civilizations accepted as a part of everyday life. Thus, preparations for life after death, through the creation of elaborate mausoleums and tombs furnished with nonliving servants for the afterlife, proved common throughout the ancient world. Many civilizations, ranging from the Americas to the Nile delta, from the Qin Dynasty in China to the Pacific islands of Polynesia, created unmoving statues that were expected to function in the spirit world and serve their creators after death.

Ancient civilizations also created extremely sophisticated machines, many with automatic functions that roughly approximated the function of modern robots. Although the primary power systems remained muscle and water, with limited reliance on wind, until the 19th century, inventors still developed complex systems. For example, the famed gardens of Babylon included automatic theaters for the entertainment of the users where water-powered machines moved and simulated living organisms. More than 2,000 years ago, Greek inventors developed automatons that relied on pneumatic power for their movement (Brooks 2002, 13). Recently, primitive examples of both a rocket engine and a steam engine were discovered near the ruins of Alexandria, suggesting that ancient inventors were far more advanced than previously assumed. In the early 13th century, Abu al-Jazari published detailed instructions for the construction of practical automata in his *Book of Knowledge of Ingenious Mechanical Devices*. While artifacts constructed according to the instructions in the book have not survived, modern engineers have agreed that the designs would function.

Automata

Although ancient societies developed substantial numbers of automata, these rarely amounted to more than a form of entertainment until the modern era. The wealthiest members of society often became interested in these entertainments and engaged the services of inventors to produce such machines. Leonardo da Vinci, widely regarded as one of the most prolific inventors of all time, designed several machines with similarity to modern military robots. The most well-known exemplar, the automatic knight, probably served as a novelty rather than having a military purpose. No working model of da Vinci's design is known to exist, although plans for its construction have been found. Modern examiners have found that the knight, as drawn, would not function, but da Vinci is known to have deliberately altered his drawings to prevent theft by his competitors. Eyewitness accounts of the knight reported that it could move its arms, rotate its head, and possibly approximate limited speech (Rosheim 2006, 112; Singer 2009, 44–45).

Later, inventors that focused on the development of automata included Jacques de Vaucanson, who built musical automata and his famous Digesting Duck for the amusement of Parisian high society in the mid-18th century (Riskin 2003). In the process, de Vaucanson also invented the first rubber tubing, thus influencing a host of later nonrobotic inventions. Friederich von Knauss also created mechanical musicians, as well as an automaton capable of writing short phrases. Pierre Jaquet-Droz built programmable android automatons that emulated human professions, and by allowing various programs to be run on his machines, some theorists have argued that Jaquet-Droz also created the first computer. At the turn of the 19th century, Henri Maillardet built realistic-looking human figures that could write and draw, utilizing a cam-based memory system for the movement of the internal parts.

Power Systems and Precedents

While de Vaucanson, von Knauss, Jaquet-Droz, and Maillardet focused on automata that simulated humans, other innovators turned their talents to the development of new mathematical systems that eventually contributed to the production of modern computers. Many of these mathematicians also sought to develop practical computation machines, capable of performing mathematical functions of ever-increasing complexity. In the mid-17th century, Blaise Pascal built a calculator that could perform addition and subtraction. Before the end of the century, Gottfried Wilhelm von Leibniz had improved Pascal's design sufficiently that the machine could also perform multiplication and division. Each of these men also contributed substantially the development of entirely new fields in mathematics.

By the 19th century, inventors could draw upon a much larger number of power sources. Internal combustion engines provided driving power to vehicles, particularly ships and railroad locomotives. Steam power drove industrial applications throughout Europe and North America. Electricity, generated in a wide variety of fashions, made the continual operation of factories possible and soon drove the industrial machinery in addition to lighting the workspace. These power sources gave inspiration to an enormous number of inventors, who responded with a host of new technological developments. In 1805, Joseph Marie Charles Jacquard developed a new punch card system to program mechanical looms. This allowed the mass production of complex woven designs and was one of the key enablers of the Industrial Revolution. Less than two decades later, British inventor Charles Babbage designed the Difference Engine, a machine that could perform far more complex mathematical functions than earlier calculators, including logarithmic and trigonometric functions. Some have argued that Babbage's next system, the Analytical Engine, first described in 1837, began the development process for modern computers.

Over the course of the 19th century, the most important technological innovation center in the world gradually shifted to the United States. In particular, communication technology emerged as a major field of innovation, made necessary by the vast expanses of the North American continent. In 1847, Samuel Morse patented his version of the telegraph system, which relied on a code to transmit signals over a single wire, rather than the far more costly multiwire systems developed in Europe. This revolutionized the speed of communication across the continent, which particularly affected the ability of military commanders to improve their situational awareness of the effects of strategic decisions. The telegraph also enhanced the role played by political leadership in American conflicts. Less than three decades later, Alexander Graham Bell demonstrated his telephone, and this was quickly followed in 1895 by the first long-distance wireless radio transmissions. All of these communication devices offered unprecedented control of enormous military forces to political leaders, who could transmit orders around the world instantaneously.

Soon, the improved communication capabilities began to affect not only the control of armies in the field and fleets at sea but also the function of individual weapons. The first significant weapons systems to utilize the new communication methods included command-detonated explosives, which could be positioned at a vital point and detonated upon the arrival of an enemy target. Torpedoes, which entered naval services in the mid-19th century, suffered from a lack of range, maneuvering, and targeting capability. Experiments with wire-guided variants proved relatively disappointing, and their usage remained largely confined to land installations guarding harbors. In 1897, Nikola Tesla demonstrated a radio-guided torpedo to the U.S. Navy. This device combined range and maneuverability into a deadly package, one that could target enemy warships at the waterline, inflicting potentially catastrophic damage at minimal risk to the user. Amazingly, the navy showed little interest in the design, perceiving it as a less-than-honorable device most suited to sneak attacks and not in keeping with the traditions of American naval service. Tesla found that many European navies had far more interest in his device's effectiveness than its ethicality, and he followed the pattern of many other American weapons inventors who created their devices in the United States but found buyers abroad.

Industrial Robotics

In the early 20th century, the United States remained the most productive center of technological innovation. Even inventors that showed little interest in developing weapons, such as the foremost American inventor of the period, Thomas Edison, still pioneered fields that eventually contributed to autonomous weapons systems. Perhaps the most important example from the period was the work of Orville and Wilbur Wright, whose *Wright Flyer* demonstrated the possibility of manned heavier-than-air flight in 1903. Less than a decade later, airplanes began military operations, first as reconnaissance craft but soon as weapons-delivery platforms. During World War I, military aviation underwent massive growth, with thousands of airplanes utilizing hundreds of designs flying in the skies over Europe. Each year, new models were introduced that made all previous models obsolete. In the last year of the war, American Charles Kettering demonstrated the Kettering Bug, the first aerial torpedo in history. The Kettering Bug could be programmed to fly a set distance before cutting off its engine, turning the aircraft into a very short-range glider. Kettering believed that his device would allow precise attacks against fortified German positions, with the Bug's 180-pound payload of explosives creating massive holes in the enemy defenses (Clark 2000, 8). With a smaller payload than that delivered by an artillery barrage, the Bug would strike enemy positions at a much greater angle of attack, making trench defenses less useful. It also had a greater range than field artillery, making it possible to strike positions in the rear of the enemy lines that had previously been immune from direct attack. In practice, none of the Bugs were deployed to Europe, and their demonstration flights did not inspire confidence from American and Allied commanders (Werrell 1985, 20).

After World War I ended in 1918, American political and industrial leaders returned their attention to increasing the domestic economy, while their European counterparts focused on rebuilding the war-torn continent. The millions killed in the war, coupled with millions more struck down by an influenza outbreak in the winter of 1918 and 1919, represented a "lost generation" of industrial workers and drove a renewed interest in machines to perform labor on behalf of humans. In 1921, Czech writer Karel Capek tapped into this sentiment with his play for English-speaking audiences titled *R.U.R.* and subtitled *Rossum's Universal Robots*. This play included the first documented use of the term "robot," a word derived from the Czech term "robota." Robota refers to the hard work traditionally performed by serf labor in Czech society. Capek's play envisioned the development of service robots that become ubiquitous in human society, performing all types of labor functions. Eventually, the self-aware mechanical servants become dissatisfied with their role and revolt against their human masters, eradicating all of humanity and becoming a new race ruling the planet. Capek's dystopian vision gave rise to an entirely new genre of literature and established a persistent theme within science fiction, namely that of machines rising up to destroy humanity (Kussi 1990, 35; Singer 2009, 66).

During the interwar period, a number of industrial robots were introduced in American factories. These machines proved successful in performing repetitive tasks, particularly when coupled with the assembly line process. Notable exemplars included William Pollard's Position Controlling Apparatus (1938) and Harold Roselund's spray-painting arm (1939), both quickly adopted by the automobile industry. This period also included a renewed interest in the creation of artificial humans, though certainly on a much more sophisticated scale than the automata of de Vaucanson and his competitors. In 1929, researchers demonstrated the first android robots to American and British audiences. Although primitive in design and function, and certainly not likely to be mistaken for humans, these machines received sensational reviews. Ten years later, the Westinghouse Electric Corporation built Elektro, a far more complex android that debuted at the New York World's Fair in 1939. The commencement of World War II derailed attempts to improve android technology, as the young field of robotics quickly converted to wartime pursuits.

World War II

While American inventors concentrated on industrial applications, their counterparts in Germany and the Soviet Union worked to develop new weapons systems that could be remotely controlled or autonomously employed. On land, the best examples of these inventions were the German Goliath and the Soviet teletank. The Goliath, a misnomer of the first order, was a small-tracked vehicle remotely controlled via a tracking wire. It could carry a payload of up to 200 pounds of explosives. In many ways, it reflected an attempt to solve the problems of the previous war rather than predict the tactical challenges of the next conflict. Had the device been available during World War I, it might have proven valuable by creeping across no-man's-land and detonating in enemy trenches. By World War II, it was too slow and undependable to contribute to the fluid warfare that characterized much of the conflict. The Soviet teletank, a radio-guided, full-sized armored vehicle, had more potential applications but was hindered by the clumsy control system, poor visibility of the operator, and lack of developed doctrine regarding proper utilization of the weapon. The few teletanks employed in the Winter War with Finland in 1939 and 1940 offered little advantage to their operators, and the cost of the system made it wholly impractical for full-scale development.

American wartime innovators returned to the results of the Kettering Bug tests of two decades earlier and sought to utilize improvements in complementary technology to massively improve upon the concept. The Bug's lack of accuracy and small payload had made it unattractive to the U.S. Army. The Army Air Forces, which had concentrated on high-altitude, precision bombing in daylight raids as their primary doctrine, had found some targets simply too heavily defended for attacks by manned bombers. Disastrous attacks such as the raid on Schweinfurt in October 1943, during which 77 of 351 attacking bombers did not survive the operation and 121 more were damaged, demonstrated that alternatives were necessary. Adding insult to injury, the raid had done negligible damage to the target, in large part due to the poor accuracy of the bombs being dropped.

To create a massive flying bomb, American engineers converted obsolete B-17 and B-24 bombers for remote operation by a trailing aircraft. Theoretically, these converted aircraft could be packed with explosives and just enough fuel to reach their target and then crashed into the enemy position. The explosives could be fitted with a timed fuse or a contact trigger that would detonate the entire load upon striking the target. Unfortunately, while the aircraft could be remotely flown once aloft, the doomed aircraft could not take off without the assistance of an onboard crew. Once airborne, the pilots, along with the flight engineer tasked with arming the explosives, were told to bail out of their moving aircraft and parachute to earth (Clark 2000, 10). Project Aphrodite, the code name for this ill-fated program, never achieved any substantial results, although Aphrodite aircraft were launched against dozens of German targets (Krishnan 2009, 18–19). The program is best known today for its most famous casualty, Joseph Kennedy, Jr., the son of Senator Joseph Kennedy and the brother of future President John F. Kennedy. He volunteered to pilot one of the experimental aircraft, and on August 12, 1944, he and his crew were killed in a premature detonation of the explosives loaded onto his airplane (Swinson 1997, 2).

Whereas American engineers had failed to build a successful flying bomb, in part due to the insistence on precision targeting, German engineers took a different approach to the issue. Allied bomber aircraft targeted specific German industrial sites for much of the war, while German attackers proved far more willing to engage in terror tactics by simply indiscriminately bombing British cities. By 1944, German engineers had successfully mated an unmanned, jet-powered aircraft and a rudimentary guidance system capable of targeting a wide area, such as London. The result was the V-1 flying bomb, a system designed to instill fear in the civilian population by randomly striking throughout the British capital. Over the next 12 months, German forces launched more than 8,000 of the devices, primarily from launching points in the Netherlands. Although the weapons failed to drive the British out of the war, they did inflict thousands of casualties and prompted British leaders to demand an aerial and ground assault to capture the launching sites and end the barrage (Werrell 1985, 61–62; Armitage 1988, 7–16).

Computers underwent major changes during and after World War II. Interest in the systems as a means to break the increasingly complex codes used by enemy forces prompted a substantial investment in their development, particularly by the Allied powers. Most of the changes were initially due to hardware improvements, as all-electronic machines began to rely on transistors rather than vacuum tubes. These modifications increased the speed and storage capacity of the computers while rapidly shrinking the size of individual machines. The emerging field of computer science also began to consider new forms of communication with computers and the programming that might utilize new computer languages. These developments began with the use of punch cards similar to those used in mechanical looms but soon moved to internally stored programs and the use of rewritable media. At the same time, industrial robots continued to proliferate. In 1951, the Atomic Energy Commission began to use the world's first teleoperated mechanical arm. This device, designed by Raymond Goertz, could be used to manipulate hazardous materials within nuclear reactors. Operators reported impressive precision from the arm's force feedback system, which allowed the arm to grip and manipulate materials with a much lighter touch. Incorporating advances in computing, George Devol designed a reprogrammable robotic arm in 1953. This design proved attractive for applications in constantly changing industries such as automobiles, as it could perform its work even after annual model redesigns.

Cold War Robotics

In 1953, the world's first remotely operated submarine began operations. Because the design did not need to include life-support apparatus, it could travel to much greater depths than manned submarines. It could also remain under water for an indefinite length of time, subject only to the limits of its battery power. This craft and its successors not only enhanced the scientific understanding of the ocean's depths but also demonstrated the potential military applications of unmanned undersea vehicles (UUVs). Some roboticists have argued that the exploration of outer space should follow the pattern of oceanography, with unmanned platforms performing most of the missions and manned systems being used only in rare circumstances, if at all.

In the 1950s and 1960s, major sources of robotic innovation began to emerge in academic and government locations. In 1958, the U.S. Department of Defense founded the Advanced Research Projects Agency (ARPA; later renamed the Defense Advanced Research Projects Agency). This institution exists to facilitate major leaps forward in technological innovation, pushing for revolutionary changes rather than evolutionary progress. It often achieves its goals by partnering with leading academic institutions or by providing seed grants to further technological research. One year after ARPA's founding, John McCarthy and Marvin Minsky cofounded the Massachusetts Institute of Technology (MIT) Artificial Intelligence Laboratory, the first such institution in the world. In 1963, McCarthy founded the Stanford Artificial Intelligence Laboratory, while Minsky continued to direct the MIT location for decades. These two academic centers remain key sources of AI research and development and have trained hundreds of researchers for the field.

In 1958, Jack St. Clair Kilby, an engineer at Texas Instruments, created the first integrated computer circuit. This revolutionized computer design and greatly increased the speed of computer miniaturization. In 1965, Gordon Moore noted that the number of components per circuit had doubled approximately every 18 months. He correctly predicted that the pace would continue, meaning the calculating power of computers would increase exponentially (Moore 1965). This concept, nicknamed Moore's Law, has been remarkably stable for more than four decades, and as a result, has been applied with mixed success to other related fields. In 1968, the Intel Corporation, founded in Santa Clara, California, began to use Kilby's circuit technique to produce microprocessors. Within three years, the company had released the world's first commercial microprocessor, and Intel remains a dominant innovator and producer in the microprocessor field.

Although the space race officially commenced in 1957 with the launch of the Soviet Union's *Sputnik*, it did not become a true robotic contest until 1966. In that year, both the Soviet Union and the United States launched robots toward the lunar surface. In a direct parallel to the earlier stages of the race to develop space capabilities, the Soviet robot, *Luna 9*, reached the moon sooner but performed with much less sophistication and for a much shorter period of time. *Luna 9* sent photographs of the lunar surface back to its earthbound controllers but functioned for only three days before falling silent. The American robot, *Surveyor*, not only photographed the moon, it also sent video images and engineering data derived from lunar samples. *Surveyor* continued to operate and report its findings for more than seven months.

In 1970, the Soviet Union sent *Lunokhod I* to the moon. Coming a year after the American manned lunar mission, *Apollo 11*, *Lunokhod I* garnered relatively little worldwide attention. However, as the first remotely operated vehicle to maneuver on the lunar surface, *Lunokhod I* demonstrated the feasibility and efficiency of using robotic systems for extraplanetary exploration. Just six years later, the American-built robots *Viking I* and *II* landed on the surface of Mars. They sampled the Martian soil and transmitted the results of their experiences for years, beginning a long series of robotic missions to Mars carried out by the U.S. National Aeronautics and Space Administration (NASA). In 1977, NASA launched *Voyager I* and *II*, a pair of robotic probes headed toward the outer planets of the solar system. Both probes continue on their outward journey today, and each is beyond the orbit of the farthest planet but still transmitting data back to Earth and responding to commands from NASA engineers.

At the end of the 1960s, robotics researchers began to focus much of their efforts on biomimetics, elements of robots that are inspired by biological examples from nature. For example, Marvin Minsky, still ensconced in the MIT labs, developed a tentacle robot arm that mimicked the limb of an octopus. It was a far cry from the stiff movements of most robotic arms, and it inspired a host of new developments in robotic manipulators. Just one year later, Stanford's Victor Sheinman created the Stanford Arm. This arm could follow arbitrary paths in space, rather than being forced upon a predetermined line, because it incorporated multi-axis articulation. Industrial buyers flocked to the new system, as it could perform multiple operations on a single target from a variety of directions, greatly increasing its value to assembly line production.

Within the realm of military robotics, leaders of the U.S. Air Force began to recognize the potential value of unmanned platforms as reusable reconnaissance devices. In particular, the Teledyne Ryan Firebee, an aerial target drone used to train interceptor pilots and air defense personnel, was modified to follow preprogrammed flight paths while carrying a camera (Wagner and Sloan 1992, 11). Soon renamed the Lightning Bug, this reconnaissance drone eventually flew more than 3,400 missions in the skies over Vietnam. Quickly recognizing the potential of the system, Teledyne Ryan engineers created models with electronic jamming capabilities that could precede manned aircraft over a target area and protect them from surface-to-air missile sites. By 1972, the rechristened Firefly could be fitted with a television camera and transmission system, and rather than flying preplanned routes, it could be remotely flown from a controller aircraft that remained in safe airspace, away from enemy ground defenses (Clark 2000, 12–15). This new arrangement allowed rapid-response missions, loitering over areas of particular interest, and up-to-the-minute battle damage analyses of areas hit by other aircraft or artillery. While the system was not weaponized, it was only a small step to move from the modified target drone to an unmanned attack platform. This could be accomplished either by adding a warhead and turning the Lightning Bug into a flying torpedo (later redesignated a cruise missile) or by adding a release mechanism that would allow the Lightning Bug to be flown on multiple missions, dropping its munitions at a preprogrammed location. Only the end of the Vietnam War prevented experiments in these directions from being operationally tested in the field.

Robots in Popular Culture

Robots became increasingly popular aspects of American culture throughout the Cold War era. Initially, they were typically portrayed as terrifying weapons, bent on destruction and immune to human emotions, such as the killer robot Gort from *The Day the Earth Stood Still* (1951). Over time, though, they gradually became more common as useful servants to human heroes, although the robot-as-villain theme never disappeared from science fiction. While the ubiquitous television series *Star Trek* (1966–1969) rarely involved any robotic components, the equally influential *Star Wars* (1977) included two independent "droids" as pivotal heroes. By the 1980s, robots had become simply another staple of entertainment media, ranging from the terrifying future vision of assassin robots conquering humanity to harmless household helpers making life a bit more convenient for their human owners. Children played with robotic toys such as Transformers, which spawned a massive television and movie franchise. Even *Star Trek* returned to the fray, with its *Next Generation* (1987–1994) series featuring an incredibly advanced android character, Data, played by Brent Spiner. Over the course of the series, Data grew to increasingly resemble his human comrades not only in appearance, but also behavior.

Although audiences became much more comfortable with the notion of robots capable of independent action, the popular cultural conception of robots designed specifically for military use became at once more realistic and far darker. One of the pervasive themes of American science fiction films is the fear of a robotic takeover of the world. Given the theme of Karl Capek's original play, it is unsurprising that this meme has continued for nearly a century. In 1984, director James Cameron released the film that has become the new archetype of the humans-against-robots concept, *The Terminator*. Although the film primarily concerned the efforts of a single android sent back through time to kill the film's protagonist, its main characters described a future in which a military computer, Skynet, seized control of the U.S. military network and commenced a global effort to destroy humanity. Arnold Schwarzenegger's portrayal of the title robot terrified audiences, encouraging the idea that autonomous technology might represent a greater threat to its creators than to the enemies it is designed to counter. The film has spawned a franchise of sequels, television programs, and assorted merchandise. Its vision of an apocalyptic future in which humanity lives in fear of machines has influenced a host of subsequent films, most notably the *Matrix* trilogy of 1999 through 2003.

Just as robots became a much more common aspect of popular culture, computers became a major aspect of everyday life. In 1976, Steven Jobs and Stephen Wozniak founded Apple Computer Corporation, a company that initially specialized in user-friendly home computers but eventually branched out into all forms of digital entertainment, communication, and information media. By 2012, Apple had become the most valuable technology company in the world (Sarno 2012). In 1981, IBM released its first personal computer, an affordable home system that made it possible for average citizens to experiment with computer programming. This system included an entirely new operating system written by a relatively unknown programmer, Bill Gates. MS-DOS soon became the dominant computer operating system, and its success enabled Gates to found Microsoft, a company that then undertook a much more ambitious project, the creation of Windows. Windows made computers far more accessible and attractive to lay users, opening the door to massive growth in the home computer industry and familiarizing subsequent generations of American students with the basics of computer function, design, and potential.

The rapid expansion of computing power into the hands of home users as well as academic institutions spurred a correlated growth of robotics researchers. For the first time, it was possible for individuals and small university departments to experiment with robotic designs. While the existing centers of robotics and artificial intelligence remained in the vanguard of research and design, particularly of large or extremely complex designs, smaller and simpler machines began to proliferate, further stimulating interest in the developing field.

Academic Centers

At the top research institutions, designers renewed their focus on the problem of how robots perceive and interact with their environment. In 1984, researchers at Tokyo's Waseda University built Wabot-2, a humanoid robot focused on musical performances. It used artificial eyes to read musical scores and performed them on an organ. Unlike de Vaucanson's musical robots of two centuries earlier, Wabot-2 proved able to read and perform completely new compositions that it had not encountered before. Naturally, the robot performed scores placed before it without regard to the resulting melodies—it had no judgment regarding the quality of the music it performed. Researchers at Stanford's Artificial Intelligence Laboratory built Flakey, a mobile robot whose stereovision eyes could recognize individual humans. The robot could then be ordered to follow specific people while autonomously navigating the crowded laboratory environment. Flakey represented a major advance in the creation of pattern-recognizing programs and served as an inspiration to many subsequent machines on weaponized systems, which are often called upon to compare objects in their visible environment to existing patterns for friendly or enemy forces. This paved the way for autonomous weapons capable of independent lethal decision making, and taken to its logical extreme, the possibility of a pro-grammable robotic assassin that could rely upon facial-recognition software to stalk its target prior to launching an attack.

In 1989, Rodney Brooks and the MIT Robotics Institute demonstrated Genghis, a small, insect-like robot that can be adapted to function in almost any environment. Unlike the more complex robots being released by other institutions, Genghis was deliberately designed to be simple and incapable of indepen-dent, sophisticated behavior. Rather, Genghis used bottom-up processing, responding continually to its immediate environment rather than attempting to map out its surroundings. Taken alone, Genghis seems to simply wander, moving to avoid obstacles but doing little, if anything, to suggest a systematic examination of its environs. When placed with multiple replicas of itself and a simple command to avoid other Genghis robots and continue to explore, the small machine demonstrated a significantly different approach to robotic modeling of environments (Baggesen 2005, 3, 67). A Genghis swarm could quickly move through a very large area, and when each returned information about its surroundings to a central processor, Brooks was able to demonstrate that biomimetic swarm behavior represented a viable alternative to individual robots of increasingly complex, expensive, and often fragile design.

Not all autonomous robots need to make life-and-death deci-sions regarding the employment of weaponry, but all need sophisticated environmental recognition systems to navigate and avoid tragic accidents. If robots could be created to be perfectly capable of travelling through a changing environment without human intervention, the long-imagined possibility of driverless vehicles could become a reality. In 1984, Defense Advanced Research Projects Agency (DARPA) sought to encourage research in this field through the creation of the Autonomous Land Vehicles (ALV) initiative (Lehner 1989, 156–159). The ultimate goal of the program was the creation of independent transport vehicles for military use, a development that would ease logistical burdens, particularly in combat situations. Within two years, designers at Carnegie Mellon's Robotics Institute created a robot that could autonomously drive a truck, albeit in a very controlled environment. Further efforts to create robotic transports have yielded a number of prototypes, including the MULE and BigDog squad-level support vehicles.

In 1988, Danny Hillis unveiled the Connection Machine, a revolutionary new approach to central processor design. Demonstrating the idea that Moore's Law regarding microprocessor design would yield applications in other aspects of computing power, Hillis devised a method for his processors to operate in parallel, allowing an unprecedented 65,000 simultaneous computations. Not only did this represent a massive increase over the speed of existing designs, it also suggested an entirely new approach to designing increasingly capable computers. Within a decade, IBM built a massively improved parallel processor. Deep Blue, a supercomputer designed with the singular purpose of playing chess, engaged in a very public match with world chess champion Garry Kasparov in 1997 and defeated him in a six-game match. It was the first time a machine had beaten one of the world's top players in a traditional match setting.

While earlier industrial robots had focused on performing repetitive tasks on assembly lines, in the 1990s, robots began to appear in far more sensitive and less expected sectors of society. In 1990, Howard Paul and William Bargar debuted RoboDoc, a precision surgical system that has subsequently revolutionized the medical profession. RoboDoc allows a surgeon's hand movements to be replicated on a minute level, which means the surgeon can insert instruments into a small incision in a patient and perform surgery using a microcamera and a television monitor. After animal testing showed enormous promise, RoboDoc began to be used with human patients in 1992. Its surgical procedures involve shorter recovery periods, lower possibilities of infections, and fewer complications. In 2001, Dr. Jacques Marescaux and Dr. Michel Gagner performed the first telesurgery using the Zeus robotic system. The surgeons remained in New York while operating on a patient located in France, performing their work through a telepresence robotic system guided by their movements.

Robotic systems based on biological designs became increasingly common in the 1990s. Researchers argued that nature had already benefitted from millions of years of evolutionary trials and had thus developed extremely complex but efficient systems that could be applied to robotic designs. To Rodney Brooks, this meant robotic cognition could be modeled on human behavior. In 1993, he began work on Cog, a robot created to incorporate new information in the same way that human children acquire knowledge. Because Cog is a machine, it is not limited in how often or for how long it can learn new things, as it never becomes tired, bored, or frustrated. As such, it has proven capable of acquiring new information at an exponential pace. MIT's Cynthia Breazeal began to examine how robots could recognize, respond to, and emulate human social cues. In 1998, she started construction on Kismet, a robot with enough physical facial characteristics to nonverbally simulate emotions that could be easily recognized by human research subjects.

In 1995, Waseda University researchers finished Hadaly-1, a humanoid robot capable of social interaction with humans and designed as a prototype social robot. The following year, Honda Motors Company completed Prototype 2, a humanoid machine that walked, climbed stairs, and picked up objects. These robots represented major advances in human-robot compatibility, an area of particular interest to Japanese researchers. In 1997, NEC Corporation began the PaPeRo, the first of a special new class of anthropomorphized machines. These "remote presence" robots are part communication systems and part teleoperated machines. A remote operator of the robot can interact directly with other humans near the machine, not only communicating but also engaging in other activities to create a companion robot driven by a distant human.

David Barnett, a researcher at MIT, demonstrated his RoboTuna in 1996. This undersea explorer's propulsion mimics a swimming fish, resulting in extremely efficient, almost silent operation. Naval acquisitions experts quickly realized the potential military applications of such a system. It could prove nearly undetectable by conventional means, as it presents a small sonar return and gives off almost none of the traditional mechanical noises needed by acoustic detection methods. In 1999, Sony Corporation released the AIBO, a robotic dog intended for the international toy market. This artificial pet followed spoken commands from its owners, gradually learning to be an obedient companion on a much faster schedule, and with greater certainty of success, than its organic forebears.

The 1990s included a renewed interest in using robots for space exploration. In 1997, Sojourner, a semiautonomous rover, landed on the surface of Mars and began to explore the Martian environment. It chose its own paths around obstacles, making it capable of much faster and safer movement than a remotely driven machine could hope for, given the fact that radio signals require between three and 22 minutes to travel between the two planets, depending on their relative positions in the orbits around the sun. Such a lag time could certainly prove fatal to a remotely driven vehicle, particularly if any interference blocked a vital command. The Sojourner's controllers expected a week of functionality but were stunned when it continued to maneuver and transmit data for three months. The same year, NASA engineers began work on Robonaut, a humanoid created to assist during spacewalk operations. Eventually, NASA administrators expect Robonaut or its successors to assume primary responsibility for these dangerous operations (NASA 2012).

Fielding Military Robots

In 1994, the U.S. Air Force (USAF) examined one of the most important unmanned systems in its history. USAF leadership remained skeptical of the utility of unmanned platforms but agreed to test the usefulness of the General Atomics Predator as an intelligence, surveillance, and reconnaissance (ISR) platform. The following year, it was accepted into service as the RQ-1, with the "R" signifying a reconnaissance aircraft and the "Q" indicating an unmanned vehicle. The same year, the Predator made its first combat deployment, with a handful of vehicles sent to the Balkans for ISR usage. Although the system experienced some problems, its initial successes demonstrated the virtually limitless possibilities of UAVs.

With the advent of the 21st century, military forces throughout the world began to commit significant resources to unmanned capabilities. To emphasize the importance to the American military in 2000, the U.S. Congress passed a bill requiring one-third of all deep strike aircraft to be unmanned by 2010, and one-third of all ground combat vehicles to be unmanned by 2015. Given the fact that in that year, the U.S. military possessed no deep strike or ground combat unmanned vehicles, the legislation set a daunting task for the military services. Assisting in the acquisition of such forces were the RQ-8 Fire Scout, first tested in 2000, and the RQ-4 Global Hawk, which completed an autonomous nonstop flight across the Pacific in 2001. That year, tests of the feasibility of arming Predators resulted in the MQ-1, with the "M" indicating a version of the aircraft armed with air-to-ground missiles. General Atomics also tested a larger version, the RQ-9/ MQ-9 Reaper, which has a much larger carrying capacity in its armed variants.

The terror attacks of September 11, 2001, ensured that American military forces would be deployed around the globe and would take with them a growing number of unmanned vehicles. In the immediate aftermath of the attacks, a variety of robots were used in search and rescue efforts at the World Trade Center. Within a few weeks, UAVs were flying reconnaissance missions over Afghanistan, and soon armed variants were prowling the vicinity seeking to attack high-value targets. The early successes of these aircraft, operating in a highly permissive environment against an enemy with virtually no air defense network, demonstrated that under certain circumstances, unmanned platforms have potentially more utility than their manned counterparts. As observation platforms, along with the aircraft's long loiter capabilities and slower flight speeds, meant a much greater capability to employ advanced sensors against ground targets. Because these aircraft operated at 25,000 feet, they were virtually undetectable at night and fairly innocuous even in daylight.

In 2002, the MQ-1 began operations in the skies over Afghanistan. Wedding a small air-to-ground missile to the previously unarmed observation aircraft allowed a nearly instantaneous response to targets that could be identified on the ground, solving a problem that had plagued early aerial operations against Taliban and Al Qaeda forces. The fleeting nature of the targets meant that the lag time between detection, confirmation, and delivery of a manned strike aircraft often allowed the target to escape. Jet aircraft proved unsuitable to the ISR mission, as the task of flying a complex fighter jet is too demanding to permit careful scanning of the terrain for enemy infantry forces. Further, the larger and louder jet aircraft offered ample warning to enemy units, who could consequently seek cover before drawing fire. Predators armed with Hellfire missiles, in comparison, could often approach undetected, identify targets, track their movements, and fire on them before the enemy even realized hostile vehicles were in the vicinity. In 2002, an experimental Predator armed with Stinger antiaircraft missiles flew a combat patrol over southern Iraq and engaged an Iraqi Mig-25 in the first UAV aerial combat in history. Although the Predator fired its missile first, it missed the enemy jet, which in turn shot down the UAV (Hume 2007, 2).

When in 2003 the United States led a coalition of nations bent upon regime change in Iraq, unmanned platforms played a vital role in the necessary ISR missions, as well as in efforts to suppress Iraqi air defenses. Although the initial invasion went extremely well, with coalition forces entering Baghdad in a matter of weeks, the occupation period soon devolved into a bitter insurgency, with American and allied forces subjected to incessant harassing attacks and roadside improvised explosive devices (IEDs). Given the lack of an air defense threat, UAVs proved extremely successful in maintaining an overwatch position above Iraqi cities, detecting the positioning of IEDs and engaging militants with Hellfire missiles.

Research into robotics at academic and government institutions continues to accelerate in the 21st century. In 2004, the Centibots Project, funded by DARPA, demonstrated that 100 small autonomous robots could engage in distributed robotics. The robots deployed themselves to collectively engage in a task, with coordinated mapping of their environment and continual updating of new information to a centralized command center. This experiment served primarily to prove the concept of swarming robotics, and it identified plenty of areas that would require further research before such a collection became a deployable asset. However, when expanded onto a larger scale, the ramifications of the idea are staggering—an army of miniature robots could potentially map and respond to threats in a small city in an amazingly short period of time. In the same year, DARPA sponsored its first Grand Challenge competition. The agency called for robotics teams to supply a vehicle that could navigate a racecourse more than 100 miles long, through the Mojave Desert's harsh terrain, without human intervention, in under eight hours. Unsurprisingly, none of the entrants finished the course. The Carnegie Mellon University Red Team had the best run yet completed only seven miles of the course, leaving the $1 million prize unclaimed. The following year, DARPA doubled the prize money but also made the course even more grueling. Five teams completed the course, including the victor, Stanford University's Stanley (Singer 2009, 135–138). Most of the 43 teams involved in the second race spent more than the prize money to build their entries, demonstrating that the competition provoked enormous investments in robotic research (Singer 2009, 265–266). In 2007, DARPA upped the stakes for its Grand Challenge by requiring robotic entrants to navigate an urban setting, to include following all traffic laws and avoiding collisions. Carnegie Mellon's Boss returned the team to the top of the competition, completing the 60-mile course in four hours.

In 2005, Wakamaru, a Japanese companion robot produced by Mitsubishi, entered the market. Japan has one of the oldest populations in the world, and thus many Japanese companies hope to create lifelike assistance robots that will help with the care of citizens with disabilities and older adults. In the same year, Cornell University robotics developers unveiled a robot that can replicate itself if provided the necessary components for such an operation. In 2006, researchers at the Stanford Artificial Intelligence Laboratory demonstrated the Starfish, a robot that can relearn how to walk after losing one or more limbs and that has proven capable of crossing virtually any terrain. There exist obvious potential applications for battlefield robots that might be able to assemble themselves and survive catastrophic damage that would render previous models inoperable.

By 2006, the wars in Iraq and Afghanistan had become bitterly divisive affairs in American politics. Five years of searching had failed to produce Al Qaeda's leader, Osama bin Laden, and American casualty counts continued to rise. Several nations had withdrawn their forces from the coalition, and within the United States, an increasingly large segment of the population pressured the Bush administration to bring the military forces home. Increasingly, American military forces came to rely on robotic platforms, including thousands of unmanned ground vehicles like the MarcBot, PackBot, and TALON, as well as short-range tactical observation aircraft like the RQ-11 Raven. In 2007, the Army tested and then deployed three Foster-Miller TALON robots that had been upgraded to carry weaponry, including machine guns and rocket launchers.

The deployment of unmanned aircraft, in comparison to their manned counterparts, receives little criticism, in part because the aircraft can be flown from well outside the combat zone, and thus no American personnel are put at risk, and thus the Department of Defense and the Central Intelligence Agency quietly began to obtain more platforms and launch more strikes. In 2007, Predators reached a collective total of 250,000 hours of combat flight time.

Just two years later, the total had doubled, and in 2010, General Atomics announced that Predators had passed the one-million-hours mark. Drone flights had contributed to the tracking and eventual killing of Abu Musab al-Zarqawi, the leader of the Al Qaeda subsidiary in Iraq. Despite official protests from the Pakistani parliament, drone strikes in the lawless Afghanistan-Pakistan border region continued to rise. In 2011, an RQ-170 Sentinel provided real-time imagery of a raid by U.S. Navy SEALs that killed bin Laden at a private compound in Abbottabad, Pakistan. Four months later, the most prominent member of Al Qaeda's Yemeni affiliate, Anwar al-Awlaki, was killed in a Predator strike.

The low-risk use of UAVs represents a potential problem for American administrations, as there is always a temptation to gather intelligence on hostile regimes under the assumption that no American lives will be at risk. When Libyan activists triggered a rebellion against Moammar Gaddhafi, European and American political leaders mulled direct intervention. While President Barack Obama chose not to commit American personnel to the region, American drones soon began to appear over the skies of Libya. Near the end of 2011, reports emerged from the Islamic Republic of Iran that the Iranian military had somehow captured an RQ-170 Sentinel, a stealthy ISR aircraft that had not been publicly confirmed to exist by the Department of Defense. At first, the U.S. government refused to acknowledge the loss of an aircraft, but it soon was forced to admit that the Iranians did appear to be in possession of such a model, which was largely intact. In 2012, the Iranian government announced it had success-fully extracted the information contained within the Sentinel, and that they had commenced building replicas of the aircraft for their own use. While it is doubtful that Iran has the capability to produce such a model, it may choose to offer the captured UAV to China or Russia for reverse-engineering, something well within the capabilities of either nation. Despite the embarrassment asso-ciated with the loss of the aircraft, the United States remains fairly aggressive with the use of unmanned aircraft, dispatching Preda-tors to Uganda to assist that nation's search for Joseph Kony, the founder of the terrorist Lord's Resistance Army. American UAVs continue to operate in the skies over Afghanistan, Pakistan, and Yemen, and are almost certainly collecting information in a host of other locations around the globe.

Humanity has always sought new technologies with which to more efficiently fight enemy populations, and the worldwide design explosion of military robots that is currently underway merely follows the same pattern that has existed for thousands of years. However, the innovations that characterize the robotics revolution represent a more pernicious threat than any weapons system that has been created to date, as for the first time, humanity may make the decision to deploy weapons that are capable of acting independent of their human users. While remotely piloted vehicles do not represent a potential threat to the entire international legal and ethical systems that govern typical behavior in war, autonomous weaponry, if unleashed, could create an existential threat to the very existence of humanity.

While the hardware and performance of robotic systems continues to improve, it is less clear if behavioral programming is becoming better, particularly in terms of ethics. Not all judgments can inherently be reduced to a mathematical representation that lends itself to computer programming, and there is certainly no way to foresee every situation that an autonomous machine might face, particularly in a combat environment. To be fair, it is impossible to train human troops to react perfectly in every situation, and it is impossible to instill a perfect ethical behavior into military personnel. Humans, though, have the benefit of a lifetime of making instant judgments in relation to other humans, and an innate understanding of what it means to both live and to kill. Until machines can be built that can definitely understand the true value of life, it is inadvisable to grant them unfettered permission to kill, even in a perfectly designed and controlled system.

The battlefield of the future may involve combat where no humans are involved at all. Machines may be deployed by each belligerent to destroy each other, and when one side has absorbed enough punishment to be open to attack on human targets, the war may be considered over and a victor declared. Such a possibility would be a radical departure from past conflicts, though, and would require a major modification of human behavioral norms. Instead, it is far more likely that military hardware will be deployed in an attempt to strike directly at the enemy's source of capability or will to carry on a conflict, even if it means bypassing the enemy's fielded forces and striking the human population by any possible means. As the only means possessed by a weaker enemy to inflict meaningful damage against a stronger opponent, the autonomous weapons (AW) of the future may be programmed to kill humans indiscriminately or based only on geographic location. Such an attack would almost certainly ignite a series of escalations and counter-escalations, until the sole purpose of the war would be to kill the enemy population in the most efficient manner possible. Rather than starting down this slippery slope, it would be far preferable to take steps to prevent or control such technology before the temptation to unleash it becomes too powerful to resist. Unfortunately, when it comes to restraint in the pursuit of new weaponry, the record of humanity does not encourage optimism that we will not need to learn from the mistake of deploying a terrifying new weapons system against a human population.

References

Armitage, Michael. 1988. *Unmanned Aircraft*. London: Brassey's Defence Publishers.

Baggesen, Arne. 2005. *Design and Operational Aspects of Autonomous Unmanned Combat Aerial Vehicles*. Monterey, CA: Naval Postgraduate School.

Brooks, Rodney. 2002. *Flesh and Machines: How Robots Will Change Us*. New York: Pantheon Books.

Clark, Richard M. 2000. *Uninhabited Combat Aerial Vehicles: Airpower by the People, for the People, but Not with the People*. CADRE Paper No. 8. Maxwell Air Force Base, AL: Air University Press.

Clarke, Arthur C. 1973. *Profiles of the Future: An Inquiry into the Limits of the Possible* (revised ed.). New York: Harper & Row, 1973.

Hume, David B. 2007. *Integration of Weaponized Unmanned Aircraft into the Air-to-Ground System*. Maxwell Paper No. 41. Maxwell Air Force Base, AL: Air War College.

Krishnan, Armin. 2009. *Killer Robots: Legality and Ethicality of Autonomous Weapons*. Burlington, VT: Ashgate.

Kussi, Peter, ed. 1990. *Toward the Radical Center: A Karel Capek Reader*. Highland Park, NJ: Catbird Press.

Lehner, Paul E. 1989. *Artificial Intelligence and National Defense: Opportunity and Challenge*. Blue Ridge Summit, PA: TAB Books.

Moore, Gordon E. 1965. "Cramming More Components onto Integrated Circuits." *Electronics* 38, no. 8: 82–85.

National Aeronautics and Space Administration (NASA). 2012. *Robonaut.* Retrieved from http://robonaut.jsc.nasa.gov/

Perkowitz, Sidney. 2004. *Digital People: From Bionic Humans to Androids.* Washington, DC: Joseph Henry Press.

Riskin, Jessica. 2003. "The Defecating Duck, or, the Ambiguous Origins of Artificial Life." *Critical Inquiry* 29, no. 4: 599–633.

Rosheim, Mark E. 2006. *Leonardo's Lost Robots.* New York: Springer.

Sarno, David. February 29, 2012. "Apple's Market Value Tops $500 Billion." *Los Angeles Times.* Retrieved from http://articles.latimes.com/2012/feb/29/business/la-fi-apple-value-20120301.

Singer, P. W. 2009. *Wired for War.* New York: Penguin.

Sundvall, Timothy J. 2006. *Robocraft: Engineering National Security with Unmanned Aerial Vehicles.* Maxwell Air Force Base, AL: School of Advanced Air and Space Studies.

Swinson, Mark L. 1997. *Battlefield Robots for Army XXI.* Carlisle Barracks, PA: U.S. Army War College.

van Joolen, Vincent J. 2000. *Artificial Intelligence and Robotics on the Battlefields of 2020?* Carlisle, PA: Army War College.

Wagner, William and William P. Sloan. 1992. *Fireflies and Other UAVs.* Leicester, UK: Midland.

Werrell, Kenneth P. 1985. *The Evolution of the Cruise Missile.* Maxwell Air Force Base, AL: Air University Press.

Williams, Kevin W. 2004. *A Summary of Unmanned Aircraft Accident/Incident Data: Human Factors Implications.* Washington, DC: Department of Transportation and Federal Aviation Administration.

2

Problems and Controversies

While the field of military robotics is relatively new, and robotic weapons have been utilized for only a few decades, several fundamental problems have already begun to emerge. It is likely that entirely new paradigms of strategic thinking will be required to deal with the ramifications of increasing battlefield automation, but for the time being, eight significant issues stand out as worthy of examination.

Artificial Intelligence

Although the term "artificial intelligence" (AI) was not coined until 1956, creating machines that are capable of matching human minds in abstract reasoning, logical deduction, or artistic expression has long been a goal of scientists and inventors. Thus far, no machine has demonstrated "intelligence" to such a degree that a scientific consensus considers it intelligent. In part, this is because scientists working in the fields of computer programming, robotics, and human behavior cannot agree on what constitutes intelligence. Some argue that it relates to raw calculating power, communication, or ability to out-play human competitors in games with complex rules, such as chess. Others argue that AI requires a computer to demonstrate self-awareness and original concepts. The most famous test of AI is the Turing Test, named for its inventor, Alan Turing. Turing proposed that a human should provide abstract inquiries to two unknown subjects, one machine and one human, through a neutral medium such as electronic text. If the inquisitor cannot distinguish between the machine and the human respondent, the machine can be considered intelligent.

The concept of AI fundamentally revolves around the definition of intelligence. Prior to the Enlightenment period, intelligence was taken for granted in Western society as a gift bequeathed only to humans by their creator. It was seen as the key to what separated humanity from other forms of life. In the 17th century, many of the foremost philosophers of Europe, and later North America, began to consider the idea of the human mind and what it might represent. Thomas Hobbes postulated the idea that thoughts constituted a "mental discourse" that performed most rationally when following methodical rules. The Hobbesian construct made no distinction between internal thoughts and external writing, suggesting that to the human brain, the two were interchangeable. Rene Descartes made the vital distinction between thought, which he considered symbolic, and the real world, grounded in objects. He saw this as an explanation of the mind's ability to imagine not only the world as it might be but also as it could never exist, bound by the physical laws that govern reality. Through this distinction, Descartes created the modern idea of the human mind. Taken further, Descartes's dualism, completely separating the mental and the physical, also argued that reason is entirely divorced from the physical world. David Hume sought to discern the laws of the mind, believing that the human brain must follow a set of rules as rational and fixed as that of the Newtonian laws of physics. He argued that thoughts represented movements of matter within the brain but occupied too small a volume to be measured with then-existing technology.

In the 19th century, attempts to define intelligence became more technically oriented rather than being grounded in philosophy. As calculating machines became more common, some inventors realized that emulating human mental activity might be the wrong route to creating a functional AI. Instead, some began to consider how to represent logic in a mechanical form, one more accessible to machines. Essentially, this created the first programming languages, allowing communication between human and machine without requiring the machine to conform to human speech patterns, and by extension, human thought patterns. When Charles Babbage designed his Difference Engine and

Analytical Engine, he effectively became the first computer scientist. After Babbage, the development of computing power largely centered around increasing the speed and complexity of mathematical functions that could be performed by computers, and these problems dominated computer engineering through World War II.

The definition of intelligence gradually evolved in the middle of the 20th century. Eventually, it came to be defined as the ability to solve problems in an efficient and effective manner, drawing on past experience to heuristically guide the pursuit of solutions. In 1963, Marvin Minsky participated in the founding of Project MAC (Mathematics and Computation) at the Massachusetts Institute of Technology (MIT). In the same year, John McCarthy founded the Stanford Artificial Intelligence Laboratory. These two institutions soon came to dominate the AI research community within academia, and they remain major research facilities studying many aspects of AI.

While the movement to develop AI began to expand, the field of computers also grew at an exponential pace. By the end of the century, the most powerful computers in the world had far outstripped the memory capacity of the human brain. In 1997, IBM's Deep Blue defeated world chess champion Gary Kasparov in a six-game match, an achievement that many theorists believed would never be possible. Not only can computers hold and process new information, they can be directly linked to one another, making the potential information available to any given computer theoretically limitless. Today, computers have almost exceeded even the most optimistic assessments of human processing speed, and they will continue to increase in raw power for the foreseeable future. At the same time, there are many limits on computing capabilities, and the most powerful computer specimens are both enormous and exceedingly expensive. Nevertheless, computing power improvements have been accompanied by the corresponding miniaturization of components. As a result, the everyday laptop of today has far more memory storage and computing speed than the supercomputers of a few decades ago.

The rise and expansion of the robotics industry both relies on the development of computers and is a key enabler for future computers. If intelligence relies in part on the ability to perceive and respond to the environment, robots offer each of these capabilities to computers. Blending the increasingly sophisticated logic of computers with the improving mobility, flexibility, and perception of robots, it is not inconceivable that a strong computer connected to a mobile robot could soon demonstrate a true capacity for self-awareness. Such a combined system might also be capable of designing even more powerful systems, and through the robot's ability to manipulate its environment, building the new systems independent of any human agency. Such self-perception and self-replication could present an enormous problem, particularly if the computer's programming includes a prioritization for self-protection. If a self-aware robot began to perceive nearby humans as a threat to its own existence, the results could be disastrous.

In the 1942 short story "Runaround," science fiction author Isaac Asimov included three laws of robotics that guided the autonomous machines of the future. In descending order of priority, the laws stated that robots could not harm humans or allow them to be harmed, must obey orders given by humans, and could not harm themselves or allow themselves to be harmed. These concepts resonated with leaders in the emerging field of robotics, but they directly contradict the expected behaviors of military robots. After all, by definition, military robots must be prepared to inflict harm on human opponents. Otherwise, stopping robotic weaponry would require only the deployment of human troops, who might effectively become human shields. While Asimov's concept suggests a more peaceful world where robots provide assistance but do not engage in violence, it is not in keeping with the history of human behavior, particularly when the military utility of robots is becoming obvious.

In the United States, there are two key government drivers of AI research, in addition to the many academic locations that specialize in the field. It is instructive that both are under the direct control and guidance of the Department of Defense. The Defense Advanced Research Projects Agency (DARPA) has long maintained an interest in AI. While the bureaucratic configuration at DARPA has repeatedly changed over the decades, the pursuit of AI and other computing advances that might have military applications has never slackened. Not only does DARPA pursue its own research agenda, it also offers grants to stimulate the work in other areas, including the original grant needed to found Project MAC at MIT. Not surprisingly, DARPA is a key driver of robotic innovation as well. The Office of Naval Research (ONR), which oversees the Marine Corps Warfighting Laboratory, has also displayed a substantial interest in AI development. With an eye toward the naval and littoral combat of the future, the ONR provides a separate research approach from DARPA, ensuring that more technological approaches will be pursued than a single agency could oversee.

There are currently no international regulations governing the development of AI technologies, nor are there any prohibitions related to the use of AI in conventional weaponry. Given the difficulty of creating such agreements, and the limited success that laws to limit technology have had, it is likely that any effort to prohibit AI weaponry would fail. It might be possible to place limits on the types of applications that could be pursued, but rogue states might simply take advantage of the self-imposed limits that other states follow. Given the exponential growth of computing capabilities, the development of true AI may occur in the immediate future, and the use of augmented computers to design their successors will probably outstrip the ability of any human agency to halt the progress. As such, the transition from "smart" weapons to truly intelligent weapons may be only a matter of time.

Robots and Drones in Asymmetrical Wars

Asymmetrical wars are those that pit two (or more) belligerents against each other, but with significantly different war-making capabilities. Many factors can trigger an asymmetrical situation. The most common have been differences in population size, geographic and terrain advantages, and resources. However, in the 20th century and beyond, the most common cause of asymmetrical war has been a major difference in levels of military technology. While the possession of inferior technology can be overcome, or at least offset, through a greater commitment to the conflict, a larger population, or superior warfighting doctrine, the side possessing lesser technology is likely to absorb a larger number of casualties in each engagement, regardless of the outcome of the entire war.

History is replete with examples of high-technology societies defeating and destroying their low-technology opponents, but technology is not an inherent guarantee of martial victory. For each case similar to the conquest of the Aztec empire by Hernan

Cortes with only 600 horsemen wearing steel armor and carrying gunpowder weapons, there is also a case such as the victory of Vietnamese insurgents over first the French and later the United States. One need only examine World War II for a pair of poignant examples related to the role that technology can play in modern warfare. On the eastern front of the European theater, the German military possessed far more advanced weaponry than their Soviet opponents and used such technology to devastating effect, particularly in the first campaign season of 1941. The German military technology, however, proved insufficient to the task of overcoming the sheer size of the Soviet nation, and German leaders underestimated the conviction of the Soviet populace to repel the invaders. Further, the Soviet willingness to pay an enormous human cost to drive back the German forces demonstrates that at times, quantity can trump quality.

In the Pacific theater, the belligerents started with far more equivalent technology, although the United States quickly improved its weapons and vehicles, while Japanese technology remained relatively stagnant for the course of the war. By mid-1943, the United States had begun to use its technological advantage to drive back the numerically superior Japanese occupiers on island after island, and when American heavy bomber aircraft came in range of the Japanese home islands, they utilized newly developed incendiary devices to wreak havoc on the Japanese populace, bombing with impunity by the summer of 1945. In August of that year, American aircraft dropped atomic weapons on Hiroshima and Nagasaki, punctuating the end of the conflict with the ultimate demonstration of military technology.

The French, and later American, experience in Vietnam also demonstrates that a motivated population, particularly one unwilling to engage in the conventional warfare preferred by the enemy, may be able to defeat a higher-technology opponent. Two decades of armored forces, aerial bombardment by jet aircraft, and superior numbers could not bring victory to the colonial and anticommunist forces. However, the cost to the Vietnamese populace was enormous. Some estimates place the number of Vietnamese killed in the period of major American involvement at over 2 million, while over 58,000 U.S. forces were killed.

Robots and drones offer a tremendous allure to many political and military leaders. They offer the possibility of projecting power across virtually unlimited distances, without an equal threat to one's own forces and populace, assuming the enemy does not possess equal capabilities. However, this advantage has drawbacks. In Afghanistan and Iraq, where unmanned weapons systems have seen their most extensive usage to date, the very use of such machines has provided a new angle of recruitment for Al Qaeda and insurgency leaders, who see these machines as evidence that coalition members are not committed to the struggle and are unwilling to risk lives in the conflict. As such, the ironic effect has been a spreading belief that the insurgents and terrorists are willing, and need continue the fight only as a means to outlast their Western opponents.

Asymmetry often blurs the lines of what can be considered acceptable behavior in wartime. When a belligerent has no ability to counter the advantages of the enemy, he or she often seeks alternate means to inflict pain and justifies them as the only means of sharing the violence with the enemy population. Attacks on civilians, use of chemical and biological agents, and acts of terrorism all become more likely in an asymmetrical environment. In the same fashion, nations often change their approach to conflict when they cannot defeat an enemy through conventional means, with the result that the entire military situation can grow to encompass targets and techniques normally considered unacceptable in wartime.

Mass Production and Specialty Design

No nation in the world has unlimited resources, despite the enormous size of many military budgets. Technological innovation is extremely expensive, particularly when the technology is expected to survive the rigors of the modern battlefield and must be sufficiently dependable to risk life on its performance. Given these constraints, the decision to pursue an entirely new field of development is not lightly undertaken, and the direction of research is of paramount importance. The risk that a future enemy will derive substantial advantages from asymmetrical development of any weapons system can be a powerful motivator.

In the realm of autonomous weaponry, the age-old question of quantity versus quality is a key consideration. In many ways, this reflects a fundamental argument in the field of robotics as a whole. Some individuals argue that robots should incorporate as many components and capabilities as possible, both as a way to make them attractive to potential customers and as a way to provide the most functionality within a single machine. Others take a network viewpoint, arguing that each machine should be kept as simple as possible, but that their collective behavior can be complex.

The argument pitting a small number of highly complex systems against a larger number of less capable models is not new. Military technology has demonstrated effects of this debate for millennia and will continue to do so for the foreseeable future. The principle of the debate can be seen in ancient armies choosing between a large number of unarmed, poorly trained spearmen or a single heavily armored warrior equipped with a chariot. In the Middle Ages, some nations placed great faith in numbers, while others preferred to rely on a smaller but more specialized military force of elites. In the 20th century, a prominent example emerged during World War II, particularly on the eastern front of the European theater, where the titanic struggle between Nazi Germany and the Soviet Union played out.

The German military sought to overcome the massive Soviet advantage in troop strength through the incorporation of increasingly sophisticated technological innovation. One iconic example of this struggle is well documented by the armored warfare practiced by each army and the types of vehicles they employed. By 1944, Germany had designed and built some of the largest and most expensive armored vehicles in history, as exemplified by the Tiger II tank. This behemoth weighed 70 tons and included armor up to 18 centimeters thick in places. Its 88-millimeter main gun could fire a round that would penetrate the armor of any other tank in the world at that time. The Tiger II was a fearsome example of German engineering, but it was also unreliable, as many unproven technologies rushed into production tend to be. The most iconic Soviet tank, in comparison, was the 26-ton T-34. This smaller vehicle had a 76-millimeter gun, and a maximum armor thickness of only six centimeters. However, it was durable and dependable, and most importantly, far less expensive to manufacture and operate. Prior to their surrender, the Germans produced fewer than 500 Tiger IIs, while the Soviet factories churned out more than 80,000 T-34s.

A more recent example of quantity and quality can be found in the debate over interceptor fighters in the American inventory. The U.S. Air Force currently has the most sophisticated and powerful aerial fleet in the world, but strategic planners must consider the likely developments of peer competitors. Some planners argued that the United States should procure new models of existing fighter aircraft such as the F-16. These proven systems can be gradually upgraded and will remain competitive for years. Others argued that the superior capabilities of the F-22 Raptor will maintain the American aerial advantage for a much longer period. The cost of each individual F-22 is approximately $150 million, while the older F-16s could be acquired for only $30 million apiece. Until and unless the United States becomes engaged in a major conventional war, it will be nearly impossible to determine whether the decision to purchase the F-22 was a mistake.

In the case of robotic and drone systems, the debate over mass production of simple systems or the procurement of small numbers of incredibly sophisticated machines has only begun. Currently, the United States has chosen to pursue systems of each type, with prominent examples of each currently deployed to Afghanistan and Iraq. At the expensive end of the spectrum are enormous unmanned aerial vehicles (UAVs) such as the RQ-4 Global Hawk, which, at $35 million apiece, cannot remotely be considered expendable. At the opposite end of the spectrum are the comparatively simple RQ-11 Ravens, which cost only $35,000 apiece. For the price of a single Global Hawk, could 1,000 Ravens perform the same function? They certainly could not fly as high, as far, or as long as their larger cousins, but if one envisions a swarm of thousands of Ravens uploading their data into a single network, they could scan a much larger area in the same time period, even accounting for the necessary recoveries and relaunches. In addition, a single surface-to-air missile battery could destroy the entire Global Hawk data collection effort, while hundreds of Ravens could be lost and the mission would still be achieved.

The case for massive numbers of simple, inexpensive robots carries special weight in circumstances involving challenging environments that might degrade or destroy individual units. These locations might be naturally challenging, such as the environmental challenges of operating in outer space or on the ocean floor. They may be challenging due to a robust enemy defense system, such as might be found in a conventional war. Their challenges might simply be due to complexity—imagine the difficulty of mapping every room of every building in even a small town, much less a major city.

There are several primary arguments for the mass production approach. These systems can be considered expendable; thus, their operators can continue to perform the mission even while absorbing substantial losses. The smaller, simpler systems do not present as many substantial engineering challenges, and thus designs can be tested, produced, and improved on a much faster timeline. Despite the lack of sophistication in individual units, swarm behavior can be extremely complex. The dispersal of individual units means at least one is likely to be near a potential target, and if multiple units are sent to a single location, they will approach on a wide number of axes, providing more viewpoints and complicating matters for any defensive efforts.

Currently fielded exemplars of the mass production approach can be found on the ground and in the air in Afghanistan and Iraq. Small land units such as the PackBot, MARCbot, and Dragon Runner all fall into this category. In particular, the robots commonly nicknamed "throw bots" exemplify this concept. Likewise, many small tactical UAVs are expendable, even if they are expected to perform repeated missions. The RQ-11 Raven, for example, costs only $35,000, and thus if lost can be easily and cheaply replaced. Despite their low cost, though, many individual units have been flown on hundreds of missions, and the airframe is expected to last for at least 100 missions before normal wear and tear requires replacement. Also, if every small unit is to be outfitted with its own systems, rather than a few high-end exemplars under a centralized control, there will be a need for an enormous number of machines and a correspondingly large requirement for technical support (Norton 2007, 67).

While the notion of swarms of robotic systems overwhelming the enemy through sheer numbers has a certain attraction, there are also potential weaknesses of this attitude toward unmanned vehicles. First, by definition, swarming robots are going to be less intelligent and thus more susceptible to making errors. Proponents of swarm technology argue that raw numbers will overwhelm any problems associated with multiple kills on the same target, but they have not offered proof to date that the concept will not lead to entire swarms expending themselves in a simultaneous attack on one enemy. Likewise, due to the simplicity of their logic systems and their weaker sensors, swarm robots are more likely to reattack targets that have already been destroyed (Baggeson 2005, 67).

There are also substantial arguments in favor of the high-end design approach. Adding capabilities to a single system makes that machine far more flexible and able to perform more missions. While each unit may require a larger, better-trained team of operators, the overall force required to utilize the complex system is normally far smaller. Getting the financial backing to produce complex systems is often easier, in part because the complexity requires assistance from a large number of sources, each of which can provide political backing. Likewise, designers can often profit more from the production of a few high-end systems than a large number of simple ones and will probably retain the sole right to production as well. Plus, sometimes the increased capabilities are completely necessary, rather than simply desired, and for some missions, only the most expensive systems can perform the job.

Key American examples of the high-end model include the RQ-4 Global Hawk, the MQ-9 Reaper, and the current unmanned combat aerial vehicles (UCAVs) in development. The Global Hawk is by far the most sophisticated high-altitude, long-endurance UAV currently in use. It can scan and track multiple targets over enormous distances and is immune to all but the most sophisticated air defenses. The Reaper can carry a substantial volume of munitions on a scale much greater than any of the small, expendable systems, none of which can carry even a single Hellfire missile. The UCAVs may represent the best example of the argument, as they are fast, stealthy, and can carry a substantial payload—all characteristics rarely, if ever, found in simpler systems. Committing to a large purchase of high-technology machines also leads to substantial savings on a per-unit basis, faster overall procurement as the producer expands to provide the greater supply, and retention of the industrial base capability to create new high-end systems that might be lost if the market disappeared (Congressional Budget Office 2011, ix–x; Drew et al. 2005, 40–41).

As demonstrated by current American procurement and deployment strategies, each side of the high-end versus mass production argument has its strengths. Both approaches are currently being pursued, although the likely reductions in future defense budgets may require one to be pursued over the other. The U.S. Air Force is the major proponent of the sophisticated system approach, while the other services are more in favor of a large number of cheaper systems. Various designers have chosen to specialize in one or the other, with new entrants to the field more likely to start small and traditional defense companies more likely to aim for the expensive systems. Other nations will likely monitor American developments before committing to a single direction, a fact that will likely assist the United States in maintaining its current lead in military robotic and drone systems.

Gold Plating and Requirements Creep

"Gold plating" is a slang term that is often applied to the acquisitions process for high-technology weaponry. It is particularly apt, and problematic, when examining the development of robotic weapons systems. It generally means to continually add unnecessary features to a new system, potentially creating more political support for its adoption but also massively driving up the per-unit cost (and the resulting defense contractor profit). Virtually all weapons research and development projects experience cost overruns or fail to meet contractual deadlines, and robotic and drone projects are particularly notorious for both. Gold plating both drives the problem and is in turn driven by it. When a new weapon is not ready for demonstration on time, one natural tendency is to make up for the slow delivery by enhancing the system's capabilities, hence lessening the frustration with nondelivery of the product. This is not inherently gold plating, as long as the military actually desires the improved parameters. However, on many occasions, the "improvements" do nothing to enhance the ability of the item to perform adequately for the mission, or even exceptionally well in its role. The original concept of the RQ-1 Predator was for an expendable intelligence, surveillance, and reconnaissance (ISR) platform that could loiter in an area for a long duration and send helpful images back to the user. Many planners expected to keep the unit cost of each Predator around $1 million, plus the cost of the ground control station. The current system is not expendable; it is barely considered acceptable for slow attrition to erode the inventory. The primary cost increase is in sensors, not the aircraft itself, but the current per-unit cost is $5 million, with a system (four aircraft and a ground control station) costing $30 million. Due to the incorporation of armaments, the loiter time of the aircraft has been halved, meaning one Predator system is sufficient to keep a single armed aircraft on station 24 hours per day (Gertler 2012, 15).

The RQ-4 Global Hawk has the ability to provide images with 20-foot resolution while in its scan mode. This is perfectly sufficient for tracking vehicular motion. It is not necessary to provide a 1-foot resolution in the scan mode, though it is technically feasible. Not only would the necessary sensor be extremely expensive, but it would also create a corresponding increase in the amount of bandwidth required to transmit the detailed images through a satellite link to a ground control station. While a few scenarios might be envisioned in which the higher resolution would be desirable, the military has not requested the addition of such a capability, and if Northrop Grumman, the producer of the Global Hawk, were to add the function and correspondingly increase the price of each unit, it would exemplify gold plating. Another major problem with gold plating that is often overlooked is its effect on the development cycle—the Global Hawk was in development for more than 100 months, longer than the F-16 Falcon or the F-117 Nighthawk. It was not just the most expensive UAV in history, it also took the longest to build (Bone and Bolkcom 2003, 33).

Gold plating is not the only source of extremely high weapons costs. Often, many stakeholders are involved in the creation of new devices, and may influence the final product despite contributing nothing to its actual design. External actors may demand additional capabilities be added to a new system, which necessarily require significant adaptation of the original concept. Occasionally, these capabilities contribute nothing to the mission for which the new system is intended. When a design transforms from a single-purpose, dedicated machine uniquely created to fulfill one mission, to an all-purpose, multifunction system, designed to fulfill many roles at once, it is often the victim of "requirements creep."

The M2 Bradley Fighting Vehicle is often considered a prime example of requirements creep. It was initially envisioned as an armored personnel carrier (APC), a vehicle designed to transport infantry forces rapidly around the battlefield while shielding them from small arms fire and shrapnel. It was not equipped with heavy weapons or a major sensor package, as its only role was the ability to quickly shift the position of dismounted soldiers. Certain elements within the Army demanded that the Bradley operate as a scout vehicle as well as an APC, a role that required greater speed, tracks for all-terrain mobility, and a turret festooned with sensors to increase visibility. To give the Bradley the ability to provide fire support for troops in combat, infantry officers demanded the addition of machine guns and a rapid-fire cannon, further drawing the vehicle away from its original mission. These additions altered the profile of the vehicle so much that it began to resemble a light tank, a fact almost certain to provoke heavier volumes of fire, particularly from enemy tanks that might not otherwise waste a main-gun round on an APC. In response, the designers increased the Bradley's armor and added an antitank missile launcher. In the end, the Army's new infantry transport looked like a tank, moved slower than other armored forces it was designed to support, was too conspicuous to serve as a scout vehicle, and could carry only five soldiers. Not only had it succumbed to a massive case of requirements creep, but the equipment additions resulted in substantial opportunities for gold plating. The initial design was slated to cost $200,000 per unit and weigh no more than eight tons. The final version, first produced in 1981, weighs 25 tons and costs more than $3 million per vehicle. Unsurprisingly, the Bradley was the first of Creighton Abrams's "Big Five" weapons to be designated for replacement, and the Stryker Brigade Combat Team vehicle more closely resembles the original design of the vehicle that became the Bradley.

For UAVs and robots, one key resource that has proven susceptible to requirements creep is bandwidth—the airwaves are extremely crowded and it will likely only get worse. Tactical commanders have become so used to having full video feed of their area that they run the risk of information overload and paralysis, and yet they continually demand even more information. During Operation Enduring Freedom, the American expeditionary force was one-tenth the size of the force deployed during Desert Storm, yet it required eight times as much bandwidth for everyday operations. Laser communication systems might alleviate part of the problem but will also increase the cost of each system that uses them (Gertler 2012, 17; Griswold 2008, 10–13).

Special Vulnerabilities of Robotic Systems

As with all weapon systems, there are inherent vulnerabilities in unmanned platforms. These weaknesses might make remotely guided or autonomous machines merely less effective in performing their functions, or they might present the possibility that an enemy could actually use them against their operator. While developers of unmanned weapons have sought to minimize susceptibility to enemy countermeasures, their novelty makes their performance in the field less predictable, and the comparative lack of experienced operators within the military means a less instinctual ability to react to surprises. Some of the potential dangers of this type of weapon are unique to unmanned or remote systems, while others are common to virtually any new, potentially revolutionary military technology.

Military and political leaders commonly demonstrate a desire to develop and possess the most cutting-edge weaponry. The allure of technological answers to military situations is not unique to Western democracies, nor is it a new phenomenon. However, this near-obsession with technological dominance has reached its apex in the 21st-century U.S. military. Currently, the United States accounts for approximately half of all global military spending, including enormous amounts of resources devoted to maintaining a technological edge over peer competitors. In the period since World War II, and particularly after the end of the Cold War, the pursuit of high-technology answers to military problems has allowed the United States to maintain a much smaller military force than its primary rivals. This has in turn retained more citizens to work in industry and related sectors of the economy that contribute to military development. The result has been an enormous research and development capability coupled with massive growth in the economy that serves to fuel the system. Robotic and drone systems are merely an outgrowth of this phenomenon.

During World War II, the United States faced an enemy in Nazi Germany that possessed more advanced high-technology systems but was overwhelmed by the enormous quantity of resources and personnel that the Allies could bring to bear. The fundamental technology lesson of World War II, that higher technology does not automatically translate into military victory, is one that the United States failed to properly appreciate and incorporate into its later military planning. Rather than concentrating on the issue of why victory had been achieved over the Axis powers, the United States turned its focus to the Soviet Union, its ally against Germany but a bitter rival in the ensuing peace. While the Soviet population roughly matched that of the United States for the next four decades, the Soviet military maintained a far larger number of personnel in uniform. To offset the enemy's numerical strength, American leaders sought technological advantages.

Less than five years after the end of World War II, American units commenced battlefield operations on the Korean peninsula. The North Korean and Chinese enemies possessed inferior technology but still managed to achieve a stalemate due to superior manpower and lower logistics requirements. In the Vietnam War, the United States had an even greater qualitative edge but could not bring all of its high-technology assets to bear against a guerrilla force that refused to fight the decisive engagements that American commanders desperately pursued. The Soviets faced a similar situation in Afghanistan, where their sophisticated aircraft, armored ground vehicles, and complicated surveillance equipment could not suppress a nomadic enemy armed primarily with bolt-action rifles.

Intelligence agencies have long sought new means to gather data from enemy targets, including the interception of various forms of electronic transmissions. While listening to enemy communication offers the tremendous advantage of potentially learning the enemy's intentions, capabilities, and troop dispositions, it also makes the listeners vulnerable to disinformation campaigns. Prior to the Battle of Midway, American cryptanalysts had cracked Japanese naval codes, and as a result, American forces knew precisely when and where to expect Japanese attacks. In comparison, a massive disinformation campaign in the months leading up to the Normandy landings, in large part carried out by false radio transmissions intercepted by the Germans, convinced Adolf Hitler that the Normandy landings were a diversion to pull defenders away from the primary target near Pas de Calais. By the time he and his military commanders recognized the deception, hundreds of thousands of Allied troops had established a beachhead on the continent.

Technological solutions cannot be allowed to substitute for good judgment, and current autonomous platforms do not possess enough artificial intelligence to assume the same roles as human actors. Deception efforts that would have no possibility of fooling a human observer can still confuse machines, taking advantage of the inherent limitations of artificial pattern recognition. Smoke, camouflage, and false heat signatures can all serve to foil automated observation systems in much the same way that similar deceptive measures can confuse humans.

Most unmanned systems are at least partially controlled by remote operators. While some sophisticated UAVs can take off and land autonomously, or even carry out their preprogrammed missions without external guidance beyond the initial orders, they still maintain data links with satellites or ground control stations, at the very least for guidance and tracking but more commonly for downloading real-time imagery for immediate tactical use. As such, the ability to send and receive signals is vitally important to the proper function of these machines and represents a fundamental vulnerability. If an enemy manages to jam or intercept the signal, it will erode or eliminate the effectiveness of the device. If an enemy spoofs the signal that an autonomous machine relies on for guidance or navigation, substituting false information in place of the truth, it may be possible to capture or destroy the machine. Such an action may have been responsible for the loss of the RQ-170 Sentinel that either landed itself or crashed largely intact in eastern Iran in December 2011.

Electronic warfare and signal jamming has existed for decades. Jamming is relatively cheap and technologically simple, often consisting of a broad-range, powerful broadcast designed to overwhelm any ability to transmit on a given frequency or a range of frequencies. Because most jamming efforts require significant power output, ground-based jammers can potentially overwhelm aerial units through the sheer size of their transmitters. With remotely operated vehicles, jamming the control signals might lead to increased operational losses, as the vehicles can crash while attempting to restore a lost signal. Jamming might also render surveillance feeds unusable or so garbled as to be virtually useless.

A more sophisticated means of using vital signals could come from signal interception. Rather than blocking the signal data, the enemy might prefer to hijack it, thus allowing them to see what the operator sees. This possibility eluded UAV designers for years, with the result that many real-time data links remained unencrypted. In 2009, it was discovered that Iraqi insurgents, using a software package available online for less than $30, had managed to intercept data feeds from American Predators. This interception might seem relatively harmless, but it allowed the insurgents to know when they were under observation, making it easier to place improvised explosive devices without being detected from the air. Planners of operations related to ambushes, troop movement, and logistics operations also benefitted from knowing precisely where enemy surveillance aircraft were operating at any given time (Jelinek 2009).

Even more threatening than jamming or signal interception is the possibility that an autonomous machine might be susceptible to cyber attack or even reprogramming. Unmanned systems are simply machines run by programs—they have neither inherent loyalty nor morality. They cannot recognize if their programming has been altered and do not have the inherent instinct to determine if a command is legitimate. Computer programs are susceptible to hostile alteration, as any victim of a computer virus can attest, and cyber warfare has grown increasingly effective in recent years. Autonomous combat systems offer the possibility to create the consummate double agent or sleeper, appearing to function normally until a critical moment, when a malfunction or loss of operator control can yield the most devastating results. Already, relatively simple robots and drones have been captured, reprogrammed, and employed by enemy forces on the battlefield. As the systems become more numerous and more deadly, the possibility of electronic defection becomes increasingly likely to occur.

Thus far, autonomous and remote systems have primarily been employed in asymmetrical conflicts. The result is that designers and operators of unmanned platforms have become complacent regarding the inherent vulnerabilities of their systems. The loss of the RQ-170 Sentinel over Iran demonstrates the danger of such an attitude. While the full details of the loss remain classified, signal loss, interruption, or interception likely played a role in the capture of the aircraft. In April 2012, Iranian authorities announced that they had successfully reverse-engineered the aircraft, and they intended to replicate it for their own use. They also began to investigate the possibility of sharing the original captured aircraft with peer competitors of the United States, such as Russia or China, who have the ability to not only replicate the existing design but also significantly improve their own UAV designs in the process. While a human operator would be unlikely to blithely land in a hostile nation, leaving his or her equipment intact, the loss of the Sentinel demonstrates how unforeseen circumstances and confusion regarding signal transmission led to a major setback in unmanned system operations.

Autonomous Robots and Lethal Force

For as long as humans have fought wars, they have sought new weapons technology to improve their chances of winning the conflict. In that sense, the development of military robots follows the established pattern of the ages. At no point, though, have the machines been given the ability to decide on their own whether to apply deadly violence to a target—there has always been a human making the final determination. The ever-improving capabilities of war machines have pushed technology to a key decision point: whether those machines should have true autonomy in the application of lethal force.

To be considered autonomous, a machine must be capable of not only performing its functions independently but also of determining which functions should be performed at all. The machine can still be given orders, but it must determine for itself how to actually fulfill them. It cannot simply follow a program or engage in automatic responses to specific stimuli. It must have a range of available responses and must select for itself how to react to its environment.

Lethal force is of particular interest in regard to autonomous machines because it crosses a major legal, psychological, and ethical threshold. Once crossed, this line will be difficult, if not impossible, to restore, making any decision to proceed monumental (Clark 2000, 73). Although the world accepts the idea of humans killing one another, at least under certain circumstances, it is not certain what the reaction to lethal autonomous machines will be. One need only look at the modern reaction to machines that kill without cognition, through simple reaction to stimuli, to see the probable reaction to "killer robots." Efforts to ban or restrict the use of land mines, which indiscriminately kill or maim anyone who triggers them, and which remain dangerous for decades, demonstrate one case of disparate populations working together to ban a class of weaponry. It is worth noting that the United States is not a signatory to the treaty banning the use of land mines, a key indicator of the possible American response to any legal efforts to ban lethal robots. Instead, the United States is pursuing truly smart mines, which will not kill indiscriminately, but will undoubtedly be highly effective (Swinson 1997, 36).

Some of the deadliest weapons in history have been indiscriminate killers, unable to differentiate between enemy forces, innocent noncombatants, and even allied units. The already-cited example of land mines is one type of such weapons, but the effects of land mines tend to be felt at the individual or small unit level. They can be used to harass, interdict, or deny movement through an area, but they do not independently move, or specifically target victims, beyond the use of magnetic or weight-based trigger systems. On a larger scale, unconventional weapons, often called weapons of mass destruction (WMDs) are considered in a separate class from other lethal devices for two reasons. They are capable of destruction on a scale far beyond conventional weapons, and once unleashed, their destructive effects are uncontrollable.

The popular view of lethal, autonomous machines is almost uniformly negative. Hollywood has produced countless science fiction films with killer robots bent on the destruction of humanity. Even the term "robot" is drawn from a work of fiction, although Karel Capek's robots were not killing machines, but rather the enslaved victims of a cruel society. Whether the robots were bent on the destruction of human society, as in the iconic 1951 film *The Day the Earth Stood Still*, or seeking to kill a single protagonist, as in 1984's *The Terminator*, these robots engendered fear in audiences, in part because they killed without emotion or remorse.

There are a substantial number of arguments in favor of developing autonomous lethal robots. One obvious position is that it will be impossible to prevent all technological development in this field. Previous attempts to limit weapon developments, including WMD nonproliferation treaties, have slowed but not stopped the expansion of such programs. Rogue states continue to develop biological, chemical, and nuclear weapons in contravention of international sanctions. While this has occasionally triggered invasion (Iraq in 2003, for example), it has more often been ignored or merely verbally condemned, with little more than poorly enforced economic sanctions against the state maintaining the weapons programs. Even though leading states have foresworn biological and chemical weapons, and moved to reduce nuclear stockpiles, they have also maintained expensive and dangerous research programs, if only to counter the weapons being developed by others. In this regard, the creation of deadly robots may follow the same pattern; even if they are banned by international law, there are simply too many nations in the world with the ability to produce these devices to ensure a complete prohibition.

From a military standpoint, autonomous killing machines offer certain attractions. While no competitor has yet demonstrated true artificial intelligence, they have been capable of outperforming humans in certain calculations for decades, and their ability to quickly process huge volumes of information in some environments is far beyond humans. They are not subject to boredom, frustration, or exhaustion, and can remain at the same alert level indefinitely, making them capable of quicker detection of threats or targets, to which they can respond more quickly than their organic counterparts. In particular, they can be extremely effective against easily defined targets. A simple order, such as "Attack anything larger than two meters that moves into range" could be used to program an extremely lethal antivehicle weapon. The weapon would still need to sense its immediate environment, determine the size and range of moving objects, and decide how to carry out its attack. Some strategists believe robotic platforms should be autonomous for specific missions, such as the suppression of enemy air defenses, as the speed of computer decision making maximizes the chances of performing the mission successfully, and thus autonomy might serve to keep manned aircraft operating in the same vicinity safer from attack (Clark 2000, 74). Of course, some UAVs cost less than a modern surface-to-air missile, so perhaps it would be more ethical to simply allow them to be shot down, in the interest of causing the enemy to expend its supply of air defenses (Baggesen, 2005, 17).

From a manpower standpoint, autonomous weaponry offers a potential solution to a thorny personnel problem. Most Western democracies have shifted to an all-volunteer military force, and despite shrinking the size of their uniformed military, some are still experiencing problems filling the ranks with qualified individuals. Any military role that can be performed by a machine frees up troops for other jobs, in the same way that industrial robots reduce human labor needs. Further bolstering this view is the point that placing robots on the battlefield could reduce the number of humans placed in danger, always an attractive idea to military and political leadership.

Military robots may serve as a deterrent against potential enemies in a way that human forces cannot. Even in an asymmetrical environment, the weaker side still has the ability to inflict some casualties on the stronger belligerent. If only machines are exposed to danger, and there is no possibility of killing humans, the weaker side in a conflict may choose to abandon their efforts or may lose morale if they are exchanging their own lives for merely damaging machines. In this regard, the popular cultural view of deadly robots may provide substantial assistance.

The use of lethal robots may allow much more precision in the application of deadly force. Theoretically, robots may reduce collateral damage and civilian casualties to extremely low, or even nonexistent, levels. If both sides of a conflict employ autonomous machines, war could become a bloodless struggle between machines, with the losing side surrendering after its robots have been destroyed or neutralized. At the very least, the use of robots on the battlefield forces a consideration of the acceptable parameters of violence. Just as the development of precision-guided munitions altered the expectations of aerial bombardment and the minimization of civilian casualties, the development of these machines may eventually trigger a net reduction in the violence and devastation of warfare.

There are significant arguments against the development and employment of autonomous, lethal machines. From a legal standpoint, these weapons have elicited substantial discussion of whether their use violates existing international law. There is an enormous body of international jurisprudence designed to mitigate the effects of conflict, and much of it governs the types of weaponry allowed. Currently, there is no universally recognized international law prohibiting the use of military robots, but some scholars have argued that the use of drones violates international agreements. Likewise, the actual employment of robots and drones may prompt their operators to direct illegal behavior through the machine that they would not perform themselves.

Even if the use of military robots is legal, many opponents argue that it is still unethical. The history of human warfare has always included at least some degree of risk for the participants, even in the most asymmetric of wars. In the case of a conflict between remotely operated or autonomous machines on one side, and humans on the other, the technologically dependent force can wage war without risk. Further, turning over decisions about lethal force to a machine raises an entirely new quandary. If the machine malfunctions and kills illegitimate targets, who should be held responsible? War crimes trials, while imperfect in application, convey at least a modicum of legal and ethical responsibility for wartime behavior. International tribunals certainly cannot be expected to prosecute military machines. Finally, while some robots have been programmed to simulate emotions, that is a far cry from understanding morals and behaving in an ethical manner. Nonliving robots cannot comprehend the finality of death in the same fashion that humans can, regardless of the limits of their programming. Therefore, there is no guarantee that a robot will apply deadly force only under ethical circumstances.

Like any other new class of weaponry, the introduction of autonomous robots to the battlefield may have substantial, unintended consequences. While the United States holds a major technological lead in this area in the early 21st century, the advantage comes at a huge cost. Every American advance demonstrates the types of machines that are feasible, making it more likely that other nations will pursue machines with similar capabilities. Granting robots the ability to determine if lethal force should be applied may be another major step to gradually relinquishing control over other aspects of human civilization, bearing out Raymond Kurzweil's prediction of a future that holds no resemblance to the past and present.

The current state of technology has the capability to turn lethal decision making over to machines, particularly under certain circumstances. As yet, no nation, including the United States, has publicly acknowledged turning over that level of control, although it is plausible that it has already happened. Some weapons, such as naval close-in weapons systems (CIWS), have a limited degree of autonomy over firing decisions but are in operation only under combat conditions of little ambiguity. With the current pace of technological development, there may be at least a short-term advantage to the first adopter of completely autonomous lethal robots. It could provide a vital advantage in the next major conflict, particularly if kept secret until a critical moment or used in a surprise attack. Like other revolutionary weapons advances, though, any advantage will likely be transitory, while the permanent effects may serve to make warfare both more destructive and more likely to erupt.

Robots and the Laws of War

For hundreds of years, legal scholars, political leaders, and international bodies have sought to regulate the conduct of state-on-state violence and hence mitigate the effects of war. A vital part of these attempts has revolved around efforts to limit who should be considered a legal combatant. Attempts to ban warfare outright have proven ludicrously ineffective, whether promulgated by religious figures, secular leaders, or interlocking defensive alliances. Agreements on acceptable conduct in war, while by no means universal, have proven far more durable and effective, though by no means perfect in their application.

Long before the development of robotic and drone technology, human societies crafted agreements limiting the legitimate applications of violence. Though often not codified into a formal system, the customs of warfare have served to constrain the behavior of military forces for thousands of years. At times, certain technologies or behaviors have been recognized as morally repugnant or threatening to the maintenance of civilization. For example, the act of piracy has long been recognized as a threat to the existence of international commerce, and thus it has been harshly punished even by states that did not fall victim to piratical acts. It is not treated as an act of war unless specifically condoned and sponsored by a state. In wartime, attacks on commerce can be perfectly legal under certain circumstances, despite their obvious resemblance to the crime of piracy. The distinction between illegal piracy and legal privateering or commerce raiding comes entirely through the development of international laws of war.

In the 17th century, Dutch jurist Hugo de Grotius codified the first modern laws of war. These ideas concerned both constraints on the resort to warfare and the limits on behavior during wartime. Not only did Grotius seek to minimize the suffering of noncombatants, he also tried to reduce unnecessary pain inflicted upon uniformed personnel. Once a combatant becomes ineffective, according to Grotius, it becomes immoral and hence illegal to inflict further harm upon the individual. Thus, it is wrong to kill or enslave prisoners of war, and enemy wounded and sick must be offered medical assistance regardless of their allegiance. Emmerich de Vattel, a Swiss philosopher, built on Grotius's writings and further sought to limit acceptable violence in warfare. By the time of his work, most of Grotius's tenets had been accepted by the most powerful nations of Europe, even though they had never been formally codified into law.

During the American Civil War, Dr. Francis Leiber, a law professor at Columbia College in New York City, created a set of regulations to govern the behavior of armies in the field. It was far more detailed than previous attempts, and at the same time, relatively easy to understand and follow. The horrors of that war, along with the experience of European belligerents in the Crimean War, prompted the 1864 formation of the International Committee of the Red Cross (ICRC), a neutral organization dedicated to ameliorating the effects of warfare. In 1899, delegates met at The Hague at the invitation of Russian Tsar Nicholas II to discuss, among other things, the laws and customs of war on land and at sea. In 1907, a second conference was held to expand and modify the agreements of the earlier convention, particularly in regard to naval warfare. These agreements banned certain classes of weapons and behavior, including exploding bullets, unanchored maritime mines, and aerial bombardments. These agreements were in effect during World War I, but due to a technicality, were not legally binding upon the participants. Some of the provisions held up, such as the prohibition of exploding bullets, while others were jettisoned immediately, including the articles forbidding the use of poisonous or asphyxiating gasses. In the aftermath of the war, the League of Nations was formed, in part to provide a forum to settle international disputes without a resort to war. In 1928, fifteen of the most powerful nations in the world signed the General Treaty for the Renunciation of War, more commonly called the Kellogg-Briand Pact. By the end of 1929, fifty-four nations had become parties to the agreement, including all three members of the Axis alliance, Germany, Italy, and Japan. Needless to say, the idea was attractive in principle but utterly failed in application (Kellogg-Briand Pact 1928).

In 1929, representatives of more than three dozen nations met in Geneva to craft a comprehensive treaty delineating the limits of acceptable wartime behavior. The resulting Geneva Conventions governed many aspects of conflict and were in force during World War II. At their heart, the Geneva Conventions held nations responsible for the behavior of their troops and restricted who should be defined as a legal combatant. Being a legal, uniformed belligerent brought both protections and responsibilities. Adherence to the conventions was mixed during World War II. There were atrocious violations on the eastern front in Europe and in the Pacific theater, but there were also extremely costly attempts to maintain both the spirit and the letter of the agreements, even by the powers that eventually lost the war.

In the aftermath of the war, the Geneva Conventions were revised to eliminate loopholes and broaden protections. The United Nations (UN) was established, replacing the ineffective League of Nations. The UN Charter specifically prohibits member states from declaring war on one another and from acting in a fashion that might threaten the peaceful coexistence of member states. In the event that two members have a dispute, they are expected to resolve it diplomatically, calling on the UN Security Council for assistance if necessary. All power to intervene militarily is vested in the Security Council, which can call on member states to provide armed forces as needed for the maintenance or restoration of peace. The implication of this agreement is that an attack on any member of the United Nations could be considered an attack on all, thus pitting an aggressor against the combined might of virtually the entire world (United Nations 1945, Chapters 6–7). Regrettably, while some nations have refrained from the legal action of declaring war on one another, conflicts have continued unabated around the globe.

The development and exploitation of weapons of mass destruction (WMDs) has led to a specific subgroup of international laws regarding biological, chemical, and nuclear weapons. The 1970 Treaty on the Non-Proliferation of Nuclear Weapons (NNPT) sought to block nations not already in possession of them from developing nuclear weapons (International

Atomic Energy Agency, 1970). Not coincidentally, every nation that held a nuclear weapons capability in 1970, and which was also a permanent member of the UN Security Council, signed and ratified the agreement, as have nearly 200 other nations. India and Pakistan refused to sign the treaty, and both have demonstrated an independent nuclear capability. Israel also never signed the treaty and is suspected of possessing nuclear weapons. North Korea and Iran signed the agreement, but North Korea withdrew in 2003, while Iran remains a formal party. The treaty has hindered nations bent on developing nuclear weapons technology, but it has not stopped rogue states from generating their own nuclear capabilities. In 1975, the Biological Weapons Convention (BWC) entered into force, and currently 177 member states have forsworn biological weapons development and deployment (United Nations 2009a). In the same manner, the 1993 Chemical Weapons Convention (CWC) called for a prohibition on the development of chemical weapons and the destruction of existing stockpiles (United Nations 2009b). While 118 states have signed this treaty, most have experienced delays in the destruction of their current weapons. Neither the BWC nor the CWC have done much to deter states from pursuing programs to develop such weapons.

As long as military robots remain remotely operated machines directly controlled by a human operator, they are legally indistinguishable from any other machine of war. Responsibility for their employment can be clearly placed upon the user, who is thus subject to the laws of war. A problem begins to emerge, however, if robots are granted autonomy on the battlefield, particularly if that autonomy extends to the use of violence rather than the simple collection of intelligence, surveillance, and reconnaissance (ISR) data. Doing so is considered by many legal scholars to be both unethical and illegal (Norton 2007, 73–74). If a robot violates one of the laws of war, who should be held responsible for the action? It could be the military commander that dispatched the machine on its mission. Of course, if there was no intent to commit a crime, it could be considered an injustice to punish the leader in question. If the violation was the result of an error in programming, perhaps the designer of the robot, or the software code writer, or the industrial producer of the machine could be held responsible. Again, though, the lack of intent to do harm would make prosecution difficult, if not impossible. Given the current state of robotic technology, the concept of punishing the robot for its illegal activity would make no sense; because robots are not people, they cannot be subjected to a legal system designed for people. None of the sentences that might be considered just for a human criminal make any sense for a machine—even the death penalty would have substantially different ramifications for a machine.

Rather than attempting to enforce the existing legal mechanisms, which in any event have proven inadequate on many occasions both during and after warfare, the victim of a violation of the laws of war might feel justified in retaliating against the perpetrator. Such retaliation might be proportionate, but it would likely also fall on noncombatants. If the original victim did not have robotic forces to match those of the transgressor, any retaliation would require humans to deliberately engage in a war crime in order to exact retribution. It can be argued that the victor in a conflict writes the history of that conflict, or at the very least determines who will be prosecuted for war crimes or crimes against humanity. Nevertheless, setting up the possibility of indiscriminate retaliation presents substantial legal and ethical problems.

Quite frankly, the current laws of war are inadequate for the likely future challenges of conflicts involving robots that possess artificial intelligence and autonomous capacity to act with deadly force. There have been calls for a new international convention to reconsider the laws of war, not solely to account for technology but also to close loopholes in existing law and account for the increasing participation in warfare by nonstate actors. Such calls have not gained significant momentum and probably will not gain much traction until the first use of autonomous, lethal machines in warfare demonstrates the need for clarification of the laws of war. In the meantime, it falls to the most technologically advanced militaries to establish behavioral precedents in the hopes that self-restraint will lead to a general consensus regarding the acceptable limits of robotic participation in war. Thus far, the United States, in particular, has trod very lightly when pushing the boundaries of robotic warfare. While the U.S. armed forces incorporate more robotic systems than any other military in the world, the actual employment of said forces has always erred on the side of caution, from a legal standpoint. The U.S. military has not granted autonomy to armed robots and has actually significantly tightened the rules of engagement governing weapons release in the use of unmanned vehicles over Iraq and Afghanistan. In part, this has been because drones offer advantages that are a luxury rather than a necessity—they make military tasks easier, safer, or more effective but do not fundamentally alter the nature of those tasks, many of which have been performed for centuries or longer. Other nations have been content to monitor American behavior and follow similar rules, for the time being. Unfortunately, it may be only a matter of time before a belligerent nation decides that it will derive substantial military benefit by breaking away from the established norms. At that time, it will be beneficial if international law has codified the acceptable usages of robots and drones in warfare, lest the legal question be trumped by practical advantage on the battlefield.

Integration of the Human-Machine Team

One of the fundamental concerns in the use of robots and drones in warfare is the consideration of how the machines will interact with their human controllers or operators. Currently, even autonomous machines do not have complete control over their own actions. In particular, lethal decisions are still made by humans, even if the machine carries out an attack in a semi-independent capacity. While computers continue to rapidly improve calculation and communication speeds, humans do not have the luxury of such upgrades. Some researchers have begun to examine the possibility of direct interfaces between humans and machines as a way to speed communication and hence improve the military capabilities of the system.

Current fielded systems are primarily controlled via wireless networks or satellite data links, although some still use fiberoptic cables trailed behind the robot. They may rely on modified laptop computers, specially designed control stations, or off-the-shelf videogame controllers. Regardless of the control system, though, they all rely on human reaction times and complicated command sequences. If a mistake is made, it can have catastrophic consequences for the system involved. Examples include a host of operator-accident UAV crashes, which are the number-one cause of unit losses of Predator systems (Haulman 2003, 6; Williams 2004, 9–11). The services differ on who should be allowed to control unmanned systems—for years, the Air Force has insisted that only rated pilots should serve as remote pilots, while the other services have all pushed to broaden the potential operator pool (Sundvall 2006, 33–34). However, imagine if the user were in direct control of the machine and could control it by thought. Further, if the machine could send information or images back to the user, not through a video feed, but by sending images directly to the brain, the effect would revolutionize not just military operations but human society as a whole.

Humans have sought to physically and mentally augment themselves in an almost infinite variety of ways. Every year, the sporting world is scandalized by allegations of chemical cheating, through the use of steroids or other performance-enhancing drugs. Likewise, some innovators have sought to push the boundaries of art and the human experience by incorporating cameras directly into their bodies. Wafaa Bilal, a professor of art at New York University, had a camera surgically implanted in the back of his head in 2010, transmitting images of everything behind him for a period of two months. Taken one step further, filmmaker Rob Spence, who lost an eye in a childhood accident, replaced his glass prosthetic with a digital recorder, sending pictures of everything he viewed to a computer. While these cameras did not directly link into the artists' brains, they did allow unprecedented insight into the daily lives of their users. Amputees now have the option of not only a cosmetic replacement limb but a potentially powered prosthetic that might even improve physical performance. The Wounded Warrior Amputee Softball Team, with a roster composed entirely of military amputees, plays an entire season of games against teams composed of fully functional players and performs relatively well against them (WWAST 2012). When these advances are coupled with teleoperated robots that can provide feedback to the user through the application of force, the initial results are staggering.

The concept of integrated organic and mechanical beings has long been a mainstay of science fiction, but it is gradually becoming a reality. In 2002, Kevin Warwick engaged in a controversial experiment. He had a microchip surgically implanted in his wrist that was designed to transmit a radio signal in response to an electrical signal in his nervous system. With a bit of practice and experimentation, Warwick was able to control a robotic hand through a wireless network, using only his thoughts. Once the connection between mind and machine was established,

Warwick's team worked to make the communication two-way, allowing sensors in the robotic hand to send pulses back into his nerves through the microchip. Eventually, his wife Irena had a simpler chip implanted, and the two were able to send pulses to one another across the room. Warwick's experiments demonstrated that direct interfaces were possible between humans and machines but did not offer much immediate practical application. Nevertheless, further experiments will no doubt test the efficacy of similar interfaces, and with a greater understanding of the human neural system, may eventually create an entire android that can be controlled by thought.

Rodney Brooks has predicted that humans will gradually incorporate robotic elements directly into their bodies. This will probably begin as efforts to improve life for individuals with disabilities, particularly people with paraplegia. He suggests that enhancements and implants will become increasingly common and hence increasingly accepted by society. As they become a part of everyday life, they will undoubtedly become a routine part of military service. Taken to its logical extreme, Brooks's argument suggests that humans will gradually become cybernetic organisms, becoming more and more like their robotic creations. In comparison, some argue that robots will entirely supplant humans on the battlefield, with a resultant atrophy of basic combat skills among human troops. Some theorists believe robots are most attractive because they cost far less than a modern soldier in the long run, and they can be simply discarded when they are broken beyond repair (Krishnan 2009, 2).

Currently, the most developed military robot technology is in the field of UAVs. These remotely piloted aircraft have significant advantages over their manned cousins, including a greater tolerance of the G-forces produced by acceleration and maneuvering, the lack of any requirement for life-support equipment, and the ability to attempt missions that would be suicidal for a live pilot. The primary reason that UAVs are become increasingly autonomous is that the speed of human input to the control unit is a major limiting factor in the operation of the aircraft. The split-second decisions needed for flight operations can be more efficiently made by the aircraft itself. Giving autonomous flight control to the machine eliminates the problems of dropped data link connections, limited sensory input for the operator, and human errors caused by fatigue, boredom, or distraction, and the possibility of paralysis due to information overload (Guetlein 2005, 6).

By bringing the interaction between humans and machines closer together, it may become possible to combine the greatest strengths of both. While a robot might be able to autonomously select the best path for flight or to traverse broken ground, it cannot make the snap judgments that are second nature to humans. A machine can distinguish between an unarmed citizen and a person carrying a rifle but cannot discern intent. Is the weapon-toting individual a hunter or a dangerous partisan? Is the nearby vehicle an enemy tank or a harmless bulldozer? Human abstract reasoning is still far in advance of computer processing, and despite the major advances in the speed of machine calculations, is likely to remain so for decades. Combining the strengths of both represents the best solution for the foreseeable future.

References

Baggesen, Arne. 2005. *Design and Operational Aspects of Autonomous Unmanned Combat Aerial Vehicles.* Monterey, CA: Naval Postgraduate School.

Bone, Elizabeth and Christopher Bolkcom. 2003. Unmanned Aerial Vehicles: Background and Issues for Congress. Washington, D.C.: Congressional Research Service.

Clark, Richard M. 2000. *Uninhabited Combat Aerial Vehicles: Airpower by the People, For the People, But Not with the People.* CADRE Paper No. 8, Maxwell Air Force Base, AL: Air University Press.

Congressional Budget Office. 2011. *Policy Options for Unmanned Aircraft Systems.* Washington, DC: Author.

Drew, John G., Russell Sharer, Kristin F. Lynch, Mahyar A. Amouzegar, and Don Snyder. 2005. *Unmanned Aerial Vehicle End-to-End Support Considerations.* Santa Monica, CA: RAND.

Gertler, Jeremiah. January 3, 2012. *U.S. Unmanned Aerial Systems.* Washington, DC: Congressional Research Service. Retrieved from http://www.fas.org/sgp/crs/natsec/R42136.pdf.

Griswold, Mary E. 2008. *Spectrum Management: Key to the Future of Unmanned Aircraft Systems?* Maxwell Paper No. 44. Maxwell Air Force Base, AL: Air University Press.

Guetlein, Mike. 2005. *Lethal Autonomous Weapons: Ethical and Doctrinal Implications*. Newport, RI: Naval War College.

Haulman, Daniel. 2003. *U.S. Unmanned Aerial Vehicles in Combat, 1991–2003*. Maxwell Air Force Base, AL: U.S. Air Force Historical Research Agency.

International Atomic Energy Agency. 1970. *Treaty on the Non-Proliferation of Nuclear Weapons*. Retrieved from http://www.iaea.org/Publications/Documents/Infcircs/Others/infcirc140.pdf.

Jelinek, Pauline. 2009. "Pentagon: Insurgents Intercepted UAV Videos" (December 17). *Army Times*. http://www.armytimes.com/news/2009/12/ap_uav_insurgents_hacked_121709/.

Kellogg-Briand Pact. 1928. Yale Law School, Lillian Goldman Law Library, Avalon Project. http://avalon.law.yale.edu/20th_century/kbpact.asp.

Krishnan, Armin. 2009. *Killer Robots: Legality and Ethicality of Autonomous Weapons*. Burlington, VT: Ashgate.

Norton, David F. 2007. *The Miltiary Adoption of Innovation*. Monterey, CA: Naval Postgraduate School.

Sundvall, Timothy J. 2006. *Robocraft: Engineering National Security with Unmanned Aerial Vehicles*. Maxwell Air Force Base, AL: School of Advanced Air and Space Studies.

Swinson, Mark L. 1997. *Battlefield Robots for Army XXI*. Carlisle Barracks, PA: U.S. Army War College.

United Nations. 1945a. *UN Charter*, Chapter 6, "Pacific Settlement of Disputes." Yale Law School, Lillian Goldman Law Library, Avalon Project. Retrieved from http://avalon.law.yale.edu/20th_century/unchart.asp.

United Nations. 1945b. *UN Charter*, Chapter 7. "Action with Respect to Threats to the Peace, Breaches of the Peace, and Acts of Aggression." Yale Law School, Lillian Goldman Law Library, Avalon Project. Retrieved from http://avalon.law.yale.edu/20th_century/unchart.asp.

United Nations. 2009a. *Biological Weapons Convention*. Retrieved from http://www.unog.ch/80256EE600585943/(httpPages)/04FBBDD6315AC720C1257180004B1B2F?OpenDocument.

United Nations. 2009b. *Chemical Weapons Convention*. Retrieved from http://www.unog.ch/80256EE600585943/(httpPages)/4F0DEF093B4860B4C1257180004B1B30?OpenDocument.

Williams, Kevin W. 2004. *A Summary of Unmanned Aircraft Accident/Incident Data: Human Factors Implications*. Washington, DC: Department of Transportation and Federal Aviation Administration.

Wounded Warrior Amputee Softball Team (WWAST). 2012. *Wounded Warrior Amputee Softball Team*. http://www.woundedwarrioramputeesoftballteam.org/.

3

Worldwide Perspective

Any examination of military robots and unmanned aerial vehicles (UAVs) must naturally commence with a study of the United States, far and away the largest producer of these cutting-edge systems. Currently, the U.S. military fields more robotic and drone platforms than the rest of the world combined, and devotes a larger amount of resources to researching and developing new systems than the next 10 largest militaries combined. However, nations around the globe have chosen to pursue unmanned and autonomous vehicles and robots for their own purposes, and dozens of countries have developed the ability to produce their own models. Dozens more have imported systems for their armed forces to operate, and possibly for their engineers to disassemble and emulate. This increasing proliferation may have radical ramifications in the coming decades and could permanently transform the nature of military conflict in the modern world.

The constantly changing nature of the field of military robotics makes any overview of global conditions subject to rapid obsolescence. Further complicating the issue, most military forces in the world do not publicize the technical specifications of their machines, making many discussions of capabilities speculative, at best. Independent developers and corporations, on the other hand, often present their new designs in many public formats, hoping to attract customers in both the foreign and domestic markets. As such, the companies producing robots and UAVs are often the best sources of information about the new state-of-the-art designs, particularly if they are in capitalist societies that do not maintain a reflexive chokehold on sources of information. The Internet has become a wonderful source of data for individuals interested in the latest robotic and remotely piloted vehicle designs, as have several independent publications and information clearinghouses, many of which are available online. Information about the key objective sources of information available to the public are contained in Chapter 8, "Print and Nonprint Resources." The current chapter is designed to offer a broad overview of the state of robotic and drone development around the world, with a sampling of the types of machines currently in development. It is not designed to be an exhaustive listing of all of the military robots and remotely piloted vehicles (RPVs) in existence today but rather a comparison of the approaches that different nations take to military technological development and a guide to the nations most committed to robotic warfare.

Europe

Outside of the United States, Europe is by far the most active region in the race to militarize robots and drones. There are many factors that drive this European trend, inlcuding geography, population demographics, political factors, and historical behaviors within the region. Currently, at least 26 European nations have active robotic and UAV programs, ranging from the broad domestic industries of the United Kingdom, France, Germany, and Italy, to the relatively recent dabbling of smaller states such as Croatia, Iceland, Luxembourg, and Portugal.

The military history of Europe stretches through thousands of years of almost constant conflict. Some authors have argued that this continual warfare spurred the development of military innovations at a far quicker pace than the rest of the globe (Diamond 1998; McNeill 1982; Parker 1988). It also triggered many social changes, including the rise of the powerful modern nation-state, an absolute necessity for harnessing the full resources of a country. The military resources of Europe spread forth in a pattern of exploration and conquest, and from the 15th century onward, Europe became the political, military, and cultural center of the world. A major factor in these conquests was the rapid adoption of constantly improving military technology, often far in advance of the native cultures being faced. One need only examine the Spanish conquest of Central and South America to appreciate the value of gunpowder and steel when confronting Neolithic opponents. Similar conquests unfolded around the globe, until European states had largely divided up the rest of the planet into colonial empires.

Of course, the European colonial system could not last on a permanent basis, as each of the major European powers sought to dominate its peers and rivals for supremacy of the continent, and by extension the globe. By the time of World War I (1914–1918), the colonial collapse was well underway, and as the European states fought one another using every improving technology, they loosened their grip on the rest of their empires. World War II (1939–1945) included some of the most rapid technological advances in human history but also guaranteed the end of European political and military hegemony, and restructured global politics into a dichotomous Cold War, with Western democracies pitting themselves against totalitarian regimes in Europe, the Soviet Union, and Asia. Technological innovation continued at breakneck speed, but the power of Europe was dwarfed by the American-Soviet rivalry. The United States and its North Atlantic Treaty Organization (NATO) brethren, in particular, turned to technology as the means to offset the "Asian hordes" of the Soviet Union's Warsaw Pact allies and the communist nations of southeast Asia. In the post–Cold War era, NATO has become the de facto military arm of the European Union, gradually adding new partners throughout Europe, including nations that border the successor state of the Soviet Union, Russia. Russia still has one of the most technologically advanced military forces in the world, but its most recent conflicts and focus have been in Asia, and as such, this country will be discussed in the Asian portion of this chapter.

United Kingdom

The United Kingdom has the broadest and most robust robotic and drone warfare program of any of the European nations. With its geographic isolation from Europe as an island nation, Britain has always relied on the power of its navy to protect the nation from invasions launched off the continental coastline. The British population has remained smaller than its peer competitors, first France and then Germany, and the island does not contain substantial natural resources. For centuries, as a result, British maritime exploration, conquest, and trade have provided the necessary resources to keep the United Kingdom a major European and global power.

Britain's technological prowess remained strong throughout the 20th century. British engineers led the world in the development of many military innovations, including aircraft, radar, and computers. British scientists played a vital role in the development of nuclear weapons, cooperating with the United States on the Manhattan Project. Britain became the third nuclear power in the world, detonating its first nuclear weapon in October 1952. British corporations such as BAE Systems, QinetiQ, SELEX, and Tasuma are among the largest and most successful robotic and drone manufacturers and exporters in the world, and the British defense industry remains a competitive and lucrative sector of the British economy.

BAE Systems, based in Lancashire, produces some of the most sophisticated robotic platforms in the world. Many of their most advanced systems, such as the Corax, Kestrel, MANTIS, Raven, and Taranis, are designed to be stealthy intelligence, surveillance, and reconnaissance (ISR) platforms capable of deep weapon strikes against a peer competitor with a first-rate air defense network. The MANTIS, named for its Multispectral Adaptive Networked Tactical Imaging System, is designed to be able to perform both extensive ISR missions and attack enemy targets. Its modular sensor system can accommodate standard sensors such as video and infrared cameras, but it can also be outfitted to gather signals intelligence or carry out electronic warfare. It also has sufficient lift capacity to deliver six GBU-12 Paveway IV laser-guided bombs, ranking it among the heaviest payload capacity armed UAVs in the world. The aircraft is capable of completely autonomous flight, including takeoff and recovery, but it can also be guided from a ground control station (GCS) using a satellite link. The BAE Taranis is a joint venture to create a stealthy unmanned combat aerial vehicle (UCAV) that includes contributions from Rolls Royce, QinetiQ, and Smiths Aerospace, with the Ministry of Defence providing the majority of the funding for the project. Like the MANTIS, this aircraft is designed to perform both long-range ISR missions and deep strikes. Unlike the MANTIS, it can carry all of its weaponry inside internal bays, increasing the stealthiness of the aircraft. Like the most advanced American UCAV models, the Taranis is described as having a cruising speed of "high subsonic," although the actual maximum flight speed is a closely held secret (BAE Systems 2011; Daly 2009, 221–227).

QinetiQ is a consortium of defense industries, founded by the Ministry of Defence in 2001. In the past 10 years, it has acquired a number of defense-related corporations, including Foster-Miller, the American producer of Talon SWORDS and other robotic systems, in 2004. As such, QinetiQ is one of the few global defense contractors that produces both military robots and drones. Prior to acquiring Foster-Miller, QinetiQ had begun developing a series of UAVs, with the most prominent models named the Observer and the Zephyr. The Observer is a tactical ISR platform with an eight-foot wingspan and a maximum launching weight of 30 kilograms. It flies at 110 kilometers per hour for up to two hours, sending data from a trio of cameras to a ground control station, where the images are combined into a single video feed. The Observer was specifically designed to allow an unskilled operator to fly the aircraft. A simple touch-screen computer allows the user to choose targets displayed in the Observer's feed, and the aircraft can then autonomously move in for closer inspection. In contrast, the Zephyr is designed as a high-altitude, long-endurance UAV. Its wingspan is nearly 12 meters, yet the entire aircraft weighs less than 30 kilograms. Its rectangular, segmented wings are covered in solar collection panels, which charge a lithium polymer batter pack that in turn drives two small electric propeller motors. This configuration allows for virtually unlimited operation, as the Zephyr can fly higher than most cloud formations and hence can almost always recharge its batteries during the daytime. The ultra-light design only allows a small payload, forcing users to choose between a digital camera or a communication relay pod (Daly 2009, 233–236; QinetiQ North America 2011).

There are a number of fixed-wing UAV producers in Britain, following the conventional designs that have proven extremely successful as ISR platforms over Afghanistan, Iraq, and Libya. The largest such developers include Tasuma, which produces a number of medium-sized observation platforms in its CSV series, and also a mini-UAV called the Hawkeye (Daly 2009, 240–242;

Tasuma UK 2009). Cyberflight's Surveillance Observation Device (SOD) series has proven popular with police agencies, as it is comprised of small, lightweight, rugged, and easily portable craft (FlightGlobal 2011). Universal Target Systems developed the Vigilant to have a longer loiter time than most mini-UAVs, theoretically allowing it to launch well in advance of any tactical requirements. Because its launch and recovery are both automatic, as are its preprogrammed flight paths using global positioning satellites (GPS) and inertial navigation, the Vigilant requires minimal ground support (Daly 2009, 244).

British developers have entered into a variety of international partnerships to enhance their robotic and drone manufacturing capabilities. QinetiQ has acquired a number of smaller companies, preferring to purchase existing technology as often as the company develops its own models. Thales UK formed a joint venture with Elbit Systems, based in Haifa, Israel, with the new company called U-TacS (UAV Tactical Systems) and based in Leicester. This company produces the Watchkeeper, a minimally modified version of Elbeit's Hermes 450 ISR and targeting platform.

British companies have also embraced some of the more radical approaches to UAV technology. FanWing was formed to develop and promote a unique design wherein a conventional fuselage, tail, and landing gear from a fixed-wing UAV are mounted to a fan wing. The fan wing has multiblade rotors that resemble an unmotorized lawn mower, and are designed to produce lift by drawing air from the leading to the trailing edges of the wing. The rotors are driven by an engine, allowing lift to be generated even when the aircraft is not moving. Testing has shown the design to produce over 18 kilograms of lift per engine horsepower, making an extremely efficient aircraft. The fan wing offers many of the advantages of both fixed-wing and rotary-wing aircraft, as it is fuel-efficient, extremely stable in flight, able to loiter at speeds as low as 24 kilometers per hour, and the fan wing design makes stalling impossible. Payload capacities for each of the designs are more than the empty weight of the vehicle, and testing suggests that the design becomes more efficient as it increases in size (Daly 2009, 230–231; FanWing 2011).

Warrior Aero-Marine of Wiltshire has created one of the few amphibious UAVs in the world. The Gull can take off from an airstrip or water surface and return to either environment to land. When flying, it can mount a variety of sensors or can be configured to undertake autonomous antisubmarine warfare missions. When operating on the water's surface, the Gull can tow on undersea unmanned vehicles (UUVs), emergency supplies, sonar arrays, or water sampling gear (Daly 2009, 244–245; Warrior Aero-Marine 2010).

British engineers have also looked to ancient technology for modern surveillance platforms. London is widely regarded as one of the most surveilled cities in the world, with an estimated 500,000 closed-circuit television cameras in operation (Palmer 2010). These cameras are mostly in the hands of private agencies, as less than 2 percent are directly funded by the government, but most are accessible to British law enforcement agencies (Davenport 2007). Some British companies have been exploring tethered aerostats, dirigibles, and unmanned airships as extremely efficient means to obtain aerial viewpoints with maximum loiter capability and minimal costs. The Allsopp Skyhook combines a helium balloon with a kite surface, optimized to remain aloft in any weather for up to two weeks on a single helium charge. They range from one to 50 cubic meters, with corresponding payload capacities of 0.3 to 25 kilograms. The BAE Systems GA22 is a radio-controlled airship that can cruise at 100 kilometers per hour for 24 hours, providing an operational radius of nearly 1300 kilometers. With its 150-kilogram lift capacity, the GA22 can carry almost any ISR payload, and offers a more stable carriage for cameras than most fixed-wing and rotary-wing aircraft (Daly 2009, 220–221).

France

France, like Britain, was once one of the world's foremost colonial empires and retains a permanent seat on the UN Security Council. France's military has long maintained a tradition of leading in technological innovations, while its political leadership has pursued a policy of independence from the actions of other global powers. In the 21st century, France remains a key member of the European Union and maintains a high standard of living for its citizenry. However, modern France also has one of the oldest populations in Europe, along with one of the highest rates of immigration, both legal and illegal. These factors have strained the French domestic budget, much of which is committed to social programs, and have placed a strain on military expenditures as well.

Robots and drones appeal to French political and military leaders as a less expensive, more dependable means to enhance military effectiveness. As such, it is unsurprising that France is home to a number of the largest robotics and UAV producers in the world. These corporations also provide substantial political appeal due to their export of unmanned systems. France is one of the world's largest producers and exporters of weaponry, ranking only behind the United States and Russia as late as 2004. By 2010, French military exports had fallen substantially, lagging behind Germany, China, and the United Kingdom (SIPRI 2011).

A research consortium in France designed the SYRANO (Système Robotisé d'Acquisition pour la Neutralisation d'Objectifs), an extremely large ISR platform closely resembling an unarmed light tank. Its extensive sensor array is well protected and particularly suited for ISR duties in urban environments. The SYRANO is the first ground robot that the French military placed into combat service, although French operators had previously developed extensive experience with UAVs in combat environments.

ALCORE Technologies offers a broad array of UAV designs for French military and law enforcement agencies. These designs range from conventional, fixed-wing, propeller-driven aircraft such as the Biodrone to tail-sitting, vertical takeoff and landing (VTOL), shrouded-rotor mini-UAVs like the Maya. ALCORE has also designed the traditional rotary-wing Easycopter, a small digital photography platform available to civilians as well as military units. The fastest ALCORE production model, the Futara, is a two-meter wingspan turbojet that can perform ISR missions or mount a fragmentation warhead for use as an expendable flying bomb (ALCORE Technologies 2011; Daly 2009, 42–44).

The European Aeronautic Defence and Space Company (EADS) produces one of France's largest UAVs, the Eagle. This long-range, strategic ISR platform resembles the Israel Aerospace Industries Heron, with an added satellite communication capability. It can carry virtually any sensor package, including a synthetic aperture radar, or it can engage in signals intelligence collection. The Eagle can also perform wide-area search and rescue missions, carrying a special emergency beacon locator to pinpoint individuals in distress. Because it can cruise at over

420 kilometers per hour and remain aloft for 24 hours, the Eagle's autonomous search function can cover an enormous area in pursuit of downed French pilots. Like ALCORE, EADS also produces unmanned helicopters, including the unmanned version of the Bruno Guimbal G2 Cabri, the Orka-1200. This helicopter can lift 150 kilograms and fly for five hours before refueling, allowing it to perform ISR, target designation, search and rescue, and electronic warfare missions (Daly 2009, 46–49; EADS 2003).

Sagem Défense Sécurité was one of the first French corporations to produce UAVs, and its aircraft have deployed with the French Army to Bosnia, Macedonia, Kosovo, and Afghanistan. The Crecerelle was designed primarily as a target designator for French artillery units, ensuring that their first round fired would strike the target with no need for corrective fires. This system was redesigned into the Sperwer (sparrowhawk) and has proven extremely popular with both the French military and international customers. Sagem engineers added a much stronger engine to increase the payload, allowing the Sperwer B to carry up to 100 kilograms, including weaponry on both internal and external hard points. Sperwer Bs are currently flown by military units belonging to Canada, Denmark, Greece, the Netherlands, and Sweden. These systems have been deployed in Bosnia (1995–1996), Kosovo (1998–1999), and Lebanon (2006) to aid in peacekeeping duties (Daly 2009, 53–58; Sagem 2011). French peacekeeping forces have also utilized American Hunters in Kosovo and during their 2003 deployment to Congo.

Like their British counterparts, French corporations have embraced radical designs for robotic and drone research. The Bertin HoverEye can take off and land vertically, using an enclosed rotary wing supporting a bullet-shaped fuselage. Once aloft, the aircraft tilts forward for cruising but can return to its vertical orientation if the operator wishes to loiter over a target. It has extremely limited range (2 kilometers) and endurance (30 minutes) but can be easily transported via backpack (Daly 2009, 44–46). The Flying Robots FR101 flies by suspending a tricycle-mounted steel vehicle with a pusher engine from a parafoil wing. While not fast, it is a very stable platform with a high payload capacity, allowing it to function as a supply drop system, particularly in humanitarian missions, and it has been used in this capacity in Africa. It can be remotely flown, preprogrammed to autonomously fly using GPS navigation, or controlled by an onboard pilot using a joystick. The parafoil wing makes a crash almost impossible, as even a complete loss of power will trigger a slow glide to the earth (Daly 2009, 51–52). The Infotron IT 180-5 mounts a spherical fuselage between two sets of rotors, making a versatile craft with both military and civil applications. Payloads can be suspended from the craft to perform ISR missions, and it is available with an electric brushless motor for nearly silent operations (Daly 2009, 52–53).

Germany

Germany has a long and proud history as a source of scientific and technological innovation. One of the earliest locations of the Industrial Revolution, by the end of the 19th century, Germany had emerged as the strongest military state in Europe and a rising economic power. Over the course of World War I, German military leaders committed to technological innovation as a means of breaking the stalemate. For the remainder of the century, German engineers continued to produce some of the most advanced military technology in the world, seeking to overcome disadvantages in geography and population size by improving the equipment of the German military.

One of the earliest attempts to create a remotely operated ground vehicle for use in war was the German Goliath. It most closely resembled a miniaturized World War I tank, with rhomboid treads surrounding a small, lightly armored box. The Goliath was designed as a single-use weapon, carrying up to 100 kilograms of explosives as a rudimentary mobile command-detonated mine. It was driven by an operator using a simple joystick connected by wires to the vehicle, with a maximum range of only 650 meters and a speed of only six miles per hour. The system remained unwieldy at best, as there was no mechanism for the operator to receive feedback from the unit, meaning that the driver had to remain able to see the Goliath as it advanced. Also, its control wires could be cut, rendering the device harmless. Because each unit cost more than 3,000 Reichsmarks (approximately US$160,000 in modern currency) to build, the Goliath simply could not be considered a cost-effective weapon. On a few occasions, Goliaths were used to disrupt enemy infantry advances, but the mobility of the Goliath in these situations was virtually irrelevant—it became little more than a glorified antipersonnel mine, and the results certainly did not justify the expenditures to produce the devices (Singer 2009, 47).

The common perception of the Third Reich's military is one of a technologically superior force, and by the end of World War II, it was in large part true. Over the course of the war, German scientists created the world's first medium-range ballistic missiles, combat aircraft using jet engines, and armored vehicles larger and more powerful than any previously designed. When Germany lost the war in 1945, scientists were engaged in nuclear, chemical, and biological weapons research; there were designs for intercontinental bombers capable of leaving the atmosphere; and engineers were developing unmanned vehicles. None of the innovations could solve the fundamental problems that confronted Germany in both world wars, though, and the combined might of Great Britain, the Soviet Union, and the United States inexorably ground down the German nation, leaving it divided and occupied for nearly five decades.

Germany possesses the largest economy and most numerous population of any European state. German engineers are world-renowned for their precision and attention to detail. As such, it is unsurprising that the German military-industrial complex has developed a thriving robotic and drone manufacturing capability. Some German research has been conducted in concert with a number of other nations, while other German companies have indigenously developed their capabilities. While the current German development and production of robots and drones is not as prolific or robust as the United Kingdom or France, it is entirely likely that Germany will overtake both in the near future and become the European leader in autonomous weaponry. Germany has already overtaken both the United Kingdom and France in weapon exports and is now exceeded only by the United States and Russia in total arms sales revenues.

EMT Ingenieur is one of German's largest producers of unmanned aircraft. The company's LUNA is a medium catapult-launched tactical UAV. The LUNA has the rare ability to turn its engine off, converting to a glider when silent operation is desired, and the engine can then be restarted in flight by the operator. The LUNA carries a video camera, infrared sensor, and miniature synthetic aperture radar simultaneously, despite having a maximum payload capacity of only five kilograms. It can fly autonomously, and its recovery, refueling, and relaunch time is less than

15 minutes. Unsurprisingly, it has proven popular with deployed units. LUNAs flew more than 600 missions in Kosovo and Macedonia. Twenty LUNAs deployed to Afghanistan in 2003 and have made a combined total of more than 4,000 flights in support of the German contribution to the International Security Assistance Force. LUNAs have also been utilized in Iraq since 2003 (Daly 2009, 65–67; EMT Penzberg, 2011).

Given the success of LUNA, EMT developed a smaller tactical UAV, the Aladin, for company-level ISR missions. This hand-launched vehicle can carry a variety of cameras that transmit real-time data feeds to the ground control station. The aircraft and its control system can fit into a backpack, and can be assembled and in flight in under five minutes. It is designed for short-range operation and cannot fly higher than 200 meters. Like most mini-UAVs, the Aladin has a short endurance, but it can fly at nearly 100 kilometers per hour, relatively fast for its size. Aladins have been sent with German army units to Afghanistan, where they flew more than 2,000 missions by 2008 (FlightGlobal, 2004). They have also been purchased by the Dutch and Norwegian militaries.

EADS Defence & Security, which has subcorporations in France, Germany, Greece, and Spain, has developed Germany's first UCAV, the Barracuda. This jet-powered, stealthy aircraft has a 7-meter wingspan and 8-meter length, making it roughly the size of the Messerschmitt Bf-109 of World War II. It incorporates a variety of sensors, including a synthetic aperture radar, and can carry weaponry in an internal bay or mounted on hard points under the wings. While it has not yet deployed with German forces, the Barracuda demonstrates that German military engineers remain among the best in the world (Daly 2009, 61–62; EADS 2009).

Rheinmetall Defense Electronics's KZO is one of the rare UAVs designed to operate in an extremely contested environment. It was envisioned as an artillery spotter, though the development process soon added other ISR functions to the KZO's capabilities. It can carry a number of modular sensors for real-time data link transmission, or it can digitally store up to 10 minutes of footage for later in-flight transmission. This function is particularly useful if the data link is temporarily lost—any useful intelligence can still be retained and still sent before the aircraft returns to base, even if it is not in real-time. The KZO can transmit even through powerful jamming, a must for a laser designation system on the modern battlefield. Because GPS signals can be disrupted, the KZO uses its sensors to compare its view with a digital map, providing a much better autonomous flight navigation than sole reliance on GPS can provide (Daly 2009, 69–71).

Italy

Italy has a long and proud military history, stretching back to the Roman Empire. Much of the power of Rome came from its engineering skills, and Roman military engineers had no equals in siegecraft and the construction of field fortifications. In the modern era, Italy's geographic position has led it to focus primarily on the Mediterranean Sea and North Africa. Italian airpower theorist Giulio Douhet established one of the earliest theories about the potential of military airpower, leading the Italian military to invest heavily in aerial innovation. For a short time during the interwar period (1919–1939), Italy possessed one of the most powerful aerial arms of any military in the world. However, the shortcomings of Italian military might were showcased in the ill-fated expedition to conquer Abyssinia, an excursion that eventually required the use of bomber aircraft with poison gasses and mechanized infantry with machine guns to overcome an enemy armed largely with primitive melee weapons.

While the Italian military has shown little interest thus far to robotics, it has adopted remotely piloted vehicles as an intrinsic part of its air force. Italian aircrews have operated American-built RQ-1 Predators over Afghanistan and Iraq, and the Italian army has obtained Pointers and Ravens from the United States for tactical use. In addition to imports, several Italian defense contractors have begun to develop indigenous UAVs (Martin 2010, 155–162). The largest producers, Alenia Aeronautica, International Aviation Supply (IAS), and Galileo Avionics, have each produced several models.

Alenia specializes in large, fixed-wing UAVs with long range and extended endurance. The BlackLynx is a military version of Alenia's Molynx ISR platform. It can remain aloft for 36 hours of autonomous operation, with a range of over 3,700 kilometers. It is likely capable of serving as a hunter-killer aircraft in the same fashion as the American MQ-9 Reaper, as it possesses external hard points and has a payload capacity of more than 600 kilograms. The company's Sky-X is a multirole aircraft with a turbojet engine that allows flight at over 640 kilometers per hour but a limited endurance of only two hours. The Sky-X has a stealthy design, including an internal bay capable of holding weapons or modular ISR gear (AleniaAeronautica 2008; Daly 2009, 126–129).

IAS also focuses on fixed-wing aircraft, although for the most part it produces much smaller models than Alenia. The IAS Raffaello is a short-range tactical UAV that has a relatively long endurance of 16 hours. It can carry digital cameras, infrared sensors, or a synthetic aperture radar, but it has a data link range of under 160 kilometers. While it is capable of autonomous flight, its navigation system is somewhat rudimentary; thus, a remote pilot is often used to ensure mission success. The Pico mini-UAV is easily portable, weighing only five kilograms, and can remain aloft for 90 minutes, a fairly long time for most tactical UAVs. The most unorthodox IAS model is the Pitagora, which has a ducted rotor propulsion system that sits in the middle of the airframe's triangular wing. It allows vertical or short takeoffs and then pivots to allow forward flight. The Pitagora can carry up to 36 kilograms of sensors for up to 16 hours of flight and uses the same ground control station, with the same range limitations, as the Raffaello (Daly 2009, 131–134; IAS 2011).

The Galileo Falco is a medium-sized, tactical ISR platform that has the capability for limited weapons carrying on wing-mounted hard points. It is designed for versatility, but like other Italian UAVs, it does not use a satellite link and thus has extremely limited range for ISR collection. Its loitering time of up to 14 hours makes it a dependable system for border and coastal monitoring missions. The Galileo Nibbeo resembles a Teledyne Ryan Firebee, with its turbojet propulsion providing a maximum speed over 1,000 kilometers per hour. It can fly pre-programmed ISR routes but must return to its data link range of 150 kilometers to transmit images. It can be equipped to precede manned strike missions, mounting jammer pods or chaff and flare dispensers to distract and confuse air defense systems (Daly 2009, 129–130; SELEX Galileo 2011).

Other European Countries

A number of European countries have begun indigenous development programs in military robotics, most by importing systems for evaluation and then enticing domestic contractors to create platforms suitable to their needs. While the United Kingdom, France, Germany, and Italy have the most robust robotic and drone programs, Ukraine, the Netherlands, Spain, Switzerland, and the Czech Republic have all created respectable efforts in the field of autonomous warfare and can be considered significant participants in their own right. What tends to separate them from the major players are the number of domestic producers involved, the low level of radical innovations, and the sophistication of their systems' capabilities to date.

Ukraine's UAV program is entirely vested with a single producer, Scientifically Industrial Systems (SIS), based in Kharkov. The Ukrainian military bases much of its organization, doctrine, and equipment on the old Soviet systems, and relying on one company for technological innovation is a hallmark of this approach. It has the advantage of concentrating researchers in a single location, which reduces competition for limited supplies and eases the problem of maintaining a secure environment for the research. However, it can also stifle creativity, and it reduces incentives for efficient or rapid production of new designs. SIS supplies a variety of fixed-wing UAVs to the military, ranging from the 10-kilogram A-3 Remez to the 750-kilogram A-10 Phoenix, an unmanned version of a pre-existing ultralight airplane. While the diverse array of sizes affect the range, ceiling, and endurance of SIS aircraft, they are generally outfitted with standardized sensors, a parachute recovery system, GPS-35 navigation, and a ground control station that uses a common CPU-686E processor (Daly 2009, 212–216).

A traditional seapower, the Netherlands has devoted substantial technological resources to improving the Dutch military fleet. Not only has the Dutch navy adopted the use of UAVs, it has also turned to autonomous robotic systems for the point defense of surface vessels. The Goalkeeper Close-In Weapons System (CIWS) is a robotic gun and tracking system capable of detecting and engaging inbound missiles and shells with limited human oversight. The Goalkeeper, which has been in service since 1980, has been licensed for international trade and is currently operated aboard naval vessels belonging to Belgium, Chile, Portugal, South Korea, and the United Kingdom.

In addition to the Goalkeeper, the Netherlands has developed a robotic cargo-delivery vehicle. EADS subsidiary Dutch Space has developed the SPADES (Small Parafoil Autonomous Delivery System). This self-guiding unit allows cargo planes, such as the Lockheed Martin C-130 Hercules, to accurately airdrop supplies while remaining up to 40 kilometers away from the target, depending on wind conditions and altitude. In testing, the SPADES has demonstrated an accuracy within 100 meters when supplies are dropped from maximum range at 9,000 meters. Beyond domestic sources, the Dutch military has enhanced its UAV profile through the purchase of Israeli Skylarks, German Aladins, and French Sperwers, all of which have been deployed to Afghanistan in support of Dutch forces (FlightGlobal 2006).

Sweden began its study of UAVs by acquiring Ugglan systems from France, but its current UAV development is centered in Linköping, at Saab Aerosystems and CybAero. Saab created a pair of advanced UCAVs, the SHARC (Swedish Highly Advanced Research Configuration) and FILUR (Flying Innovative Low-Observable Unmanned Research). These turbojet-powered aircraft each weigh approximately 60 kilograms, and can fly at nearly 320 kilometers per hour. They are built of stealthy materials and can carry weapons inside internal bays (Daly 2009, 201–203). Saab used the knowledge gained through these programs to join the European nEUROn development effort, spearheaded by Dassault. The Saab Skeldar is a helicopter UAV capable of fully autonomous operation. Like the SHARC and FILUR, it is fairly small for its mission at 95 kilograms, but this gives it an endurance of up to five hours (Saab 2010). It is based on the airframe of the CybAero APID 55 (Autonomous Probe for Industrial Data Acquisition). The APID can also operate completely autonomously, allowing it to remain in complete radio silence. It can perform ISR, electronic warfare, and search and rescue missions. The APID has been deployed in civilian and military roles, and exported to foreign militaries (CybAero 2011; Daly 2009, 200–201).

In addition to obtaining UAVs from foreign sources, most notably the Israeli Harpy, Spain is home to a branch of EADS (EADS CASA) and funds UAV development at the Instituto Nacional de Téchnica Aerospacial (INTA) in Madrid. The ALO

(Avión Ligero de Observación) entered service with the Spanish army in 2000 and looks like a miniaturized P-51 Mustang from decades earlier. This system was designed as a low-cost, short-range surveillance vehicle with video and infrared options. It remains in service but has been augmented by the acquisition of American RQ-11 Ravens and Israeli Searchers. INTA improved on the ALO with the SIVA (Sistema Integrado de Vigilancia Aérea) in 2003. The SIVA has twice the wingspan but 10 times the weight of the ALO, providing more lift for the incorporation of combined video and infrared with additional space for a synthetic aperture radar, electronic warfare packages, or additional fuel. The SIVA can fly for 10 hours with its maximum fuel load, but the line-of-sight control system, even when enhanced by antennae, limits the mission radius to less than 160 kilometers (Daly 2009, 197–199).

A fiercely neutral nation, Switzerland prides itself on self-reliance. Thus, it is not surprising that after the acquisition of Israeli IAI Scouts, the Swiss government turned to RUAG Aerospace in Emmen to produce a domestic variant. The Ranger, first delivered to the Swiss Air Force in 1998 and later exported to Finland, is designed as a one-size-fits-all ISR platform that can withstand the challenges of flying in a mountainous region. This requires it to be extremely stable in flight, able to deal with the massive thermal changes in the airstream, and capable of deicing its surfaces. Its most unique feature is a specialty skid landing system that allows the Ranger to be recovered on any semiflat surface, including icy fields or frozen lakes. RUAG Aerospace began work on a larger successor, the Super Ranger, designed to bridge the gap between tactical and medium-altitude, long-endurance UAVs, but abandoned the project when the Swiss military lost interest (RUAG 2006).

The Czech Republic has a long history as a source of military production. In the 20th century, the Škoda Works ranked as one of the largest munitions factories in Europe, and it continued to churn out weaponry during the Cold War. After the collapse of the Warsaw Pact, Czechoslovakia chose to split along ethnic lines into the Czech Republic and Slovakia, and the privatization of Škoda essentially destroyed any weapons manufacturing done at its subsidiaries. The Czech military uses systems developed by the Air Force Research Institute (VTÚL in Czech) and Track Sytem, as well as imported models. The VTÚL Sojka III entered service with the Czech military in 2001. It is a fixed-wing tactical UAV similar in size and function to the RQ-2 Pioneer developed by the United States and Israel in the 1980s. The Track System Heros is a medium-sized helicopter UAV that can perform ISR, target acquisition, search and rescue, and electronic warfare missions. Its most unique feature is an automatic collision-avoidance system that can detect objects smaller than ten centimeters and take autonomous actions to dodge them (Kenyon 2005).

Europe currently has seven nations (Austria, Bulgaria, Finland, Greece, Poland, Serbia, and Slovenia) that have developed small domestic UAV capability. Each of these nations relies entirely on a single in-country source for UAVs, although most have also imported more advanced models from other producers. Austria's Schiebel Eletronische Geräte has produced the Camcopter series of helicopter UAVs, which have proven dependable, easy-to-operate, and flexible in terms of sensor payloads. The Camcopter S-100 is one of the only UAVs in the world that can mount a ground-penetrating radar. Camcopters have been sold to at least eight foreign militaries (Genuth 2007). Bulgaria's Aviotechnica Sokol looks like a slightly modified remote-controlled model airplane. It has a video camera, but no datalink capability, and its reliance on radio control limits its range to just over three miles (Daly 2009, 14–16). Finland's Patria MASS (Modular Airborne Sensor Systems) is a 1.5-meter wingspan, electric motor-driven monoplane that can interchangeably mount a video or infrared camera, or environmental sensors. Interestingly, the Finnish Defense Forces first employed this UAV to survey the replacement of antipersonnel minefields with newer mines (Daly 2009, 41). Greece is home to a subsidiary of EADS (EADS 3 Sigma), which has designed and produced the Nearchos, a medium-sized multipurpose fixed-wing UAV. The flexibility of the modular sensor system allows the Nearchos to perform virtually any unarmed UAV task, and its endurance of up to 12 hours makes it well suited for maritime patrols along the Greek coastline. The Greek military also utilizes French Sperwers in a limited capacity. Poland's Air Force Institute of Technology in Warsaw has created the whimsically named HOB-bit, a mini-UAV for battlefield surveillance. It follows a preprogrammed flight plan and can transmit video, infrared, or digital stills for up to 45 minutes. The Polish Air Force also flies Israeli Orbiters and American Shadow 200s. Serbia's Utra Aircraft Industry produces the IBL-2004, a medium-sized, fixed-wing

UAV first designed in Yugoslavia in the 1980s but later redesigned to mimic the AAI Shadow 600. It typically uses a visual spectrum camera but can be outfitted for electronic warfare and jamming missions (AirSerbia 2005). Slovenia's Aviotech RVM04 is likewise a relatively small tactical UAV, though its real-time data link and endurance of four hours demonstrate substantially better technical capabilities.

Five European nations (Belgium, Denmark, Hungary, Norway, and Romania) operate imported systems but have done little or nothing to generate their own development programs. Belgium possesses the Northrop Grumman/IAI RQ-5 Hunter, and deployed them in Bosnia and the Democratic Republic of Congo in support of peacekeeping operations. Denmark has imported Sagem Sperwers from France and AV RQ-11B Ravens from the United States. Hungary bought two Casper 250/Sofar systems from Top I Vision in Israel in 2006 for evaluation purposes and formed a marketing alliance with the Czech Republic to develop the Sojka. Norway purchased the German EMT Aladin system in 2006. Romania purchased two AAI RQ-7 Shadow systems in 1998 and deployed them in support of operations in Iraq.

Perhaps the most sophisticated UAV currently in development in Europe is the nEUROn, a massive project to develop a stealthy UCAV for the military forces of the European Union. The cooperative effort, being coordinated by the French company Dassault, includes contributions from several European UAV producers. As the overall project leader, Dassault possesses half of both the costs and the benefits of the project. Sweden's Saab is a one-quarter stakeholder, while Italy's Alenia Aeronautica claims 22 percent of the project. The remainder is held by Spain's EADS Casa, Greece's EAB, Switzerland's RUAG, and France's Thales. The program, which made its first flight in 2012, is designed to produce a subsonic, stealthy UCAV that can penetrate modern air defense systems to deliver kinetic munitions. The total cost of the project is estimated at over 400 million Euros (Dassault Aviation 2006).

Asia

The two largest military powers in Asia, China and Russia, are unsurprisingly the earliest regional adopters of military robotics and have done the most research and development of autonomous weapons systems on the continent. Japan, long a leader in robotics design, has done little to militarize its robots, though it could undoubtedly do so in short order if desired. South Korea, Malaysia, Singapore, and Taiwan have begun to incorporate robotic systems into their armed forces, largely in the form of UAVs. India and Pakistan have created nascent programs, although neither nation seems poised to invest substantial sums into technological advancement in this field. The Philippines, Sri Lanka, and Thailand have thus far remained largely content to import a few UAV systems from international producers, although each nation has the capability to pursue independent research should it be deemed necessary or desirable. The Philippine army uses the Blue Horizon, an Israeli UAV, Sri Lanka has purchased Searcher IIs from Israel, and Thailand uses the R4E Sky Eye from the United Kingdom to test the utility of unmanned systems.

China

China currently maintains one of the largest and most diverse military research programs in the world. Advancements are pursued in semiprivate industry, government research laboratories, and academic institutions, with the result a wide variety of creative solutions to traditional military problems. Historically, the Chinese military has relied on the nation's enormous population as the primary means of creating adequate military forces. However, in the past two decades, China has gradually begun to pursue advanced technology capable of matching any potential enemy's systems. The results include stealth aircraft, antisatellite missile capabilities, a massively improved naval force, and intercontinental ballistic missile upgrades. While the seemingly inexhaustible manpower pool in China has allowed the creation of the largest army in the world, it has proven a logistical nightmare when asked to project power beyond China's borders. The pursuit of autonomous military vehicles might present a potential solution to this centuries-old problem. Although often stereotyped as a state obsessed with secrecy, China has actually demonstrated many of its latest models at air shows, designed in part to promote systems designated for export. However, details on the most advanced Chinese systems remain closely guarded, making any evaluation of the state-of-the-art designs a challenging proposition at best and an exercise in speculation at worst.

One of the largest Chinese manufacturers of UAVs is Xian ASN Technical Group, an industrial producer that works closely with the Pilotless Vehicle Research Institute, located on the campus of the Northwestern Polytechnical University. This close relationship is typical of the Chinese approach to military research and closely resembles the strategy pursued by the armed forces of the United States. Xian ASN produces a wide variety of light and medium UAVs, ranging from the Xian ASN-15, a low-cost ISR platform weighing less than seven kilograms, to the ASN-207, a medium-range multirole UAV weighing nearly 500 kilograms. The 207 can do standard ISR missions but is also capable of laser designation of targets and communication relays. It is unique because multiple 207s can be flown in tandem, with one acting as a communication relay to the other, allowing a massive expansion of the system's range. When coupled with an endurance of more than 16 hours, the result is an operational range of several hundred kilometers. If aerial refueling capability is added to the system, it becomes a formidable platform that can loiter indefinitely while maintaining a secure data link for real-time updates to ground commanders, essentially creating the same capabilities as a satellite in geosynchronous orbit (Daly 2009, 33–37; Von Kospoth 2009).

China's Nanjing Research Institute on Simulation Technique (NRIST) has partnered with Beijing University to develop a series of rotary-wing UAVs. These range from the NRIST I-Z, a nine-kilogram autogyro suitable only for short-range tactical reconnaissance, to the NRIST Z-3, a 100-kilogram remotely piloted helicopter suitable as a communication relay and surveillance system. NRIST has also developed fixed-wing UAVs with the advantage of rocket-assisted launches that can be performed without a runway. Because these NRIST systems (the W-50, PW-1, and PW-2) are all recovered via parachute landings and have an endurance of at least six hours, the result is a versatile system suitable for use in virtually any terrain or environmental condition (Daly 2009, 31–33; Von Kospoth 2009).

The Guizhou Aviation Industry Group (GAIC), with assistance from the Chengdu Aircraft Design and Research Institute, has developed some of China's most sophisticated systems, including a competitor to the U.S. Global Hawk. The GAIC Soar Dragon, a 7,500-kilogram high-altitude, long-endurance high-altitude, long-endurance (HALE) aircraft, has a range of more than 7,000 kilometers. Because it flies at nearly 18,000 meters in altitude, the Soar Dragon can fly over many antiaircraft defenses, including some aerial interceptors. Its payload capacity of at least 650 kilograms makes it capable of carrying virtually any ISR package, and it may be able to deliver kinetic weapons as well. The GAIC WZ-2000 is one of the few jet-powered UAVs in the world. Despite a wingspan of less than three meters, the WZ-2000 can fly more than 800 kilometers per hour, presumably allowing high-speed intelligence collection over well-defended ground (Daly 2009, 30–31; Strategy Page 2008).

The China Aerospace Science and Industry Corporation (CASIC) has undertaken the task of developing micro-UAVs for tactical utilization by the People's Liberation Army. Their first production models, the CASIC LT series, range from 22- to 61-centimeter wingspans, with endurance from 20 to 40 minutes. Although relatively slow, these units are ideal for missions to provide a look "over the hill" for the tactical commander. The Beijing Wisewell Avionics Science and Technology Company (BWAST) has also begun development of small UAVs, beginning with the BWAST AW series. These hand-launched, easily portable UAVs are slightly larger than the CASIC versions, allowing for an increased range, flight ceiling, and endurance at only a slight cost in convenience.

The Beijing University of Aeronautics and Astronautics worked to reverse-engineer several American reconnaissance drones shot down in the 1960s and 1970s. The result is the Chinese version of the Teledyne Ryan Firebee, named the BUAA Chang Hong. It is an air-launched UAV that follows a preprogrammed flight plan before being recovered in flight by a specially modified helicopter. Thus far, it has largely been used for scientific research, although it is also potentially useful as a reconnaissance drone or as a practice target for interceptor training missions. Other Chinese manufacturers of UAVs include the Chinese Aviation Industry Corporation and the Chengdu Aircraft Industry, each of which has created prototype models for the Chinese military. Specifications for these prototypes remain classified, and they have not been demonstrated at international air shows to date.

China's pursuit of high technology for its military forces may represent a substantial shift in the strategic mindset of Chinese leadership. The nation must confront the results of decades of enforcing a deliberately low birthrate, which has reduced the rate of population expansion but has also created a rapidly graying population unfit for military service. China may not be able to maintain its enormous military personnel base but could conceivably alleviate many personnel shortages through the incorporation of new technology. In addition, the development of a domestic robotics capability will serve to modernize the growing Chinese industrial base, potentially allowing a continuance of the massive growth rates of the Chinese economy in the early 21st century. Robotic and drone technology may also represent a deliberate attempt to keep pace with peer competitors such as the United States and Russia, while developing a substantial munitions export industry. Keeping pace in the technology race will undoubtedly deter aggression against Chinese interests, while a robust weapons export program will allow the development of client states, particularly in the resource-rich but underdeveloped portions of the globe.

Russia

The Russian Federation has maintained the legacy of the Soviet Union by attempting to remain a global power capable of direct competition with the United States. However, the transition to a democratic, capitalist system has hindered many military research projects and fragmented the formerly strong development program. The fragmentation of the Soviet Union also denied many Russian corporations access to research facilities located in newly independent states and has reduced both the population and resource base upon which the state and its corporations may draw for the production of new technologies.

Scientists and engineers of the Union of Soviet Socialist Republics (USSR) experimented with radio-controlled tanks prior to and during World War II, eventually creating remote-controlled teletanks based on five existing tank models. Because the Soviet tank designs did not keep pace with the West, Soviet planners believed that a direct tank engagement would be a losing proposition. Thus, Soviet planners wished to design a system that would allow the offensive and defensive capabilities of armored vehicles, without subjecting highly trained military personnel to enemy fire. Teletank operators could remain protected inside manned tank models held as far back from the front-line teletanks as possible, driving their remote machines at distances of up to a mile via radio signals.

Teletanks were used by Soviet forces in the Winter War (1939–1940) against Finland, hardly a first-rate opponent on par with Nazi Germany. Even against the Finns, the newer, more fluid style of warfare that characterized World War II made the ponderous teletanks fairly ineffective. Soviet prospects against the German blitzkrieg proved even more bleak, and most of the remaining Soviet teletanks were destroyed in the initial German advances, often serving only to consume a single round of German antitank ammunition. Even Soviet doctrine recognized the possible futility of teletanks—if an enemy reached and seized one of the unmanned vehicles, the operator in the manned tank had orders to destroy the teletank with his own tank's main gun, lest it be turned against fielded Soviet forces.

Many of the most important modern UAV developments in Russia have been created by preexisting aircraft manufacturers, such as the Russian Aircraft Corporation MiG (formerly Mikoyan and Gurevich Design Bureau), Mil Helicopters, Tupolev Public Stock Company, and Yak Aircraft Corporation (formerly A. S. Yakovlev Design Bureau Joint Stock Company). All four of these companies retain their headquarters and research facilities in Moscow. However, some new competitors have moved into the field, most notably A-Level Aerosystems, based in Izheusk, and the Scientific Production Corporation, Irkut Joint Stock Company, located in Irkutsk. In many ways, the new facilities are embracing more radical concepts for UAV design, while the older companies build vehicles based on existing manned aircraft.

The Russian Aircraft Corporation MiG (RAC MiG), the descendant of the most well-known Soviet producer of combat aircraft, has access to decades of aircraft design research, allowing the development of a first-rate autonomous craft capable of precision air-to-ground strikes. The MiG Skat has a stealthy design, including internal weapon bays to reduce the craft's radar signature. It can reach speeds of 800 kilometers per hour, with an operational range of over 1,900 kilometers and a payload capacity of 2,000 kilograms. In addition to conventional bombs, this UAV is outfitted to attack enemy ships and radar installations. It is approximately the same size as many existing MiG airframes, which suggests that it can take off and land on airstrips capable of servicing Russian fighter aircraft (Daly 2009, 175–176; Warfare.ru 2007).

The Mil Helicopters Corporation, like RAC MiG, has essentially converted one of its conventional systems for pilotless operation. The Mil Mi-34BP is simply an unmanned version of the Mi-34S helicopter. The newest version is very efficient, as it is able to accommodate a payload of over 500 kilograms, nearly the weight of the empty airframe. This helicopter is not purposely designed for military missions, although it is capable of performing ISR missions if needed (Daly 2009, 176–177; Hutchison 2005).

Tupolev PSC, originally known for producing large military transport and bomber aircraft, has also entered into the arena of UAV development. Given the difficulty, and lack of a market, of converting cargo aircraft to pilotless operation, beginning in the 1960s, Tupolev began to instead concentrate on creating reconnaissance drones,. Their Tu-143 and Tu-243 drones resemble the Teledyne Ryan Firebee of the same era. The Russian version proved useful for quick surveillance along a preprogrammed route, and nearly 1,000 were built, with hundreds still in service with Russian forces. Tupolev has several projects in the developmental stage, including a medium-range attack UAV, the BAK SD, and a high-altitude, long-endurance model, the Dozor-600 (Daly 2009, 179–181; Pyadushkin 2010; Tupolev PSC 2011).

Yakovlev, long a competitor to MiG in the production of fighter aircraft, has proven the most adaptable of the traditional manufactures. Its Pchela UAVs, first designed in the 1980s, are effective short-range and tactical platforms. Because it is launched from a BTR-D armored personnel carrier, the Pchela does not require an airstrip. It can be deployed in minutes, and a real-time data link allows the immediate transmission of video images. It has been used regularly during military operations against rebels in the breakaway republic of Chechnya, where it has enhanced the Russian military's ability to respond quickly to enemy troop movements (Daly 2009, 181–182; Hutchison 2005).

A-Level Aerosystems has focused almost exclusively on developing micro- and mini-UAVs. While each of their systems is capable of military usage, most have been sold to the Russian Interior Ministry or private Russian corporations with security concerns. The collapse of the USSR led to increasing problems with terror attacks, industrial sabotage, and acts of banditry, all of which are magnified by the enormous distances in the Russian interior. One of the most successful models is the ZALA 421-02, a small helicopter designed for remote sensing, pipeline inspections, and surveying. Most of the A-Level systems are hand-launched with flying wing configurations, a design that results in an efficient airframe with a long endurance relative to the size of the vehicles. They are all controlled by a common ground control station (GCS), with a touch-screen laptop to fly the aircraft and a handheld joystick to control the payload (Daly 2009, 164–168; ZALA Aero 2009).

Scientific Production Corporation Irkut JSC, like A-Level, is a recent entrant to the UAV field. Most of its airframes are fixed-wing craft, with wingspans ranging from two to five meters. However, it has also researched and tested tethered aerostats and optionally piloted vehicles, such as the Irkut-850, a fixed-wing UAV weighing nearly 900 kilograms that can undertake both ISR and aerial resupply missions. Allowing remote pilots to train by actually flying the aircraft from within the cockpit before transitioning to a ground control station creates a much greater feel for the capabilities of the aircraft and how it will perform under different conditions. However, the control system for the Irkut-850 when it is in the pilotless mode does not mirror the pilot controls in the cockpit, thus creating a certain degree of frustration for some pilots in training who have difficulty adapting to the time-lag of remote piloting and the different command mechanisms in place (Daly 2009, 168–174; Irkut Corporation 2010).

The vast geography and declining population of Russia make aerial vehicles and satellites the only practical means of maintaining a reasonable degree of border security. Further, the uninhabited reaches of Siberia beckon with untold mineral wealth, readily available to the first company able to reach and extract it. UAVs offer an obvious mechanism for exploration without the risk of piloted aircraft, and ground-based remotely operated vehicles may soon be capable of actually pulling the resources from the earth and transporting them to refinement centers.

India

Despite having the second-largest population in Asia, a correspondingly large military force, and a recent history of border clashes with Pakistan, India has been slow to develop its robotic and UAV technology, preferring to allow other nations to make the large expenditures necessary for cutting-edge research. The Indian military remains largely content to import vehicles, most notably the Israeli Harpy, Heron, and Searcher II platforms. However, Aeronautical Development Establishment in Bangalore has recently worked to create an indigenous UAV capable of multirole functions and long endurance. The Rustom, being developed exclusively for the Indian military, is designed for ISR, communication relay, and possible delivery of munitions. The airframe currently being tested can carry a payload of 350 kilograms and has a maximum endurance of greater than 24 hours (Aviation Week 2010; Daly 2009, 75–76). The mini-UAV of note being developed in India is the Kadet AeroSystems Trogon, a tactical ISR platform with a single fixed-view camera and relatively short loiter time. While the Indian military has clearly recognized the potential utility of autonomous systems, upgrading other aspects of national security has clearly taken priority over developing a domestic manufacturing capability for military robotics.

India has faced a substantial number of terrorist attacks in the past few decades, and the number has continued to grow in recent years. In particular, improvised explosive devices (IEDs) have become a significant threat, as have hostage scenarios such as the takeover of the Taj Mahal Hotel in Mumbai in November 2008. One potential response has been the incorporation of robotic technology similar to that operated by American military forces in Iraq and Afghanistan. The Indian Defence Research and Development Organization (DRDO) has indigenously produced the Daksh, a wheeled robot driven by an operator located up to 500 meters away. It can be equipped with cameras, x-ray machines, and other sensory equipment. A gripper arm can manipulate suspicious objects, which can then be disabled with either a jet of water or an integrated shotgun. It is capable of climbing stairs and navigating tight confines, and it has enough power to tow small vehicles. The Daksh can also be outfitted with electronic jamming systems to prevent remote triggering of explosive devices (Kulkarni 2008).

As a rising naval power, India has begun to acquire substantial numbers of surface vessels, including its Indian-built aircraft carrier. The survivability of such vessels is of paramount concern in the modern era, particularly given the massive investment of money and resources that their construction entails. To address this issue, the Indian navy has turned, in part, to a robotic close-in weapons system. The AK-630 consists of a radar unit synchronized with a six-barreled Gatling gun. It can track inbound missiles and shells, recognize friendly and hostile fire, and engage targets at a range of up to 4,000 meters, all at a speed far in excess of human capabilities (Indian Military 2010).

Pakistan

India's border rival, Pakistan, has a much greater commitment to UAV development. In part, this is due to recent experience as an ally of the United States in the conflict with Al Qaeda. Not only has the Pakistani military seen the potential effectiveness of autonomous vehicles on the battlefield, but there have also been hundreds of airstrikes launched on Pakistani soil by American Predator and Reaper aircraft (Roggio and Mayer 2009). Pakistani military leaders, always concerned with how to overcome India's massive personnel advantages, have turned to advanced technology as one such potential route. Military assistance funding from the United States has further enhanced Pakistan's ability to pursue advanced military technology. In 2006, the Pakistani army purchased German LUNA systems, and one year later, the air force bought Falcos from Italy. However, Pakistan has recently begun domestic design and production of UAVs.

The best developed of Pakistan's many UAV research locations is Integrated Dynamics. Based in Karachi, Integrated Dynamics has created a host of short-range surveillance UAVs, several models of which have been exported to Australia, Libya, South Korea, Spain, and the United States, in addition to purchases made by the Pakistani armed forces. The Integrated Dynamics aircraft are characterized by relatively short ranges but long loiter times, making them ideal for a variety of ISR tactical missions and border security operations. Their primary drawbacks are the need for conventional airstrips for launch and recover, and a relatively bulky guidance and control unit, although some Integrated Dynamics systems can now be controlled by a single laptop computer (Daly 2009, 157–161; Integrated Dynamics 2010).

Several other Pakistani developers of UAV technology have recently emerged, each seemingly hoping to carve out a niche serving the Pakistani military. Air Weapons Complex E-5 in Wah Cantonment carries out government-sponsored classified research and has produced the Bravo, a tactical ISR unit provided to the Pakistani army. Integrated Defence Systems in Islamabad created a comparable UAV, the Huma-1, that has a lighter payload but can provide a real-time data link to operators. National Development Complex (NDC), also in Islamabad, produced the Vector for the army, with the ability to offer a variety of quickly interchangeable sensor packages. NDC hoped to export the Vector but has found marketing difficult in the crowded field of short-range tactical vehicles (Daly 2009, 161–162).

Japan

Despite Japanese dominance in the development of new robotic systems, relatively little Japanese research has been conducted toward the production of military robots and remotely piloted vehicles. However, given the nature of military technology, some Japanese systems might fall under the restrictions applied to dual-purpose technologies. One such example comes from the Yamaha Corporation, which began testing unmanned helicopters in the 1980s. Although the R-50 and RMAX were both designed for remote spraying of agricultural areas, it requires little imagination to envision converting these UAVs to deliver chemical or biological weapons. More than 2,000 have been produced for use inside Japan, with a few dozen licensed for export, ostensibly for purely agricultural or research purposes, including nine systems sent to China in 2001 (Daly 2009, 135–137; Hanlon 2004).

South Korea

South Korea, which shares a hostile border with North Korea, has moved rapidly to develop significant robotic and UAV capabilities. It has maintained an extensive military presence along the demilitarized zone (DMZ) for almost 60 years and has come to increasingly rely on advanced technology to balance North Korea's population advantages. Although South Korea has also benefitted from substantial international assistance in defending its border, it cannot assume that other nations will permanently maintain an armed presence to guarantee its sovereignty. In addition to its troops on the border, South Korea has begun to test automated systems, including UAVs, as a potential deterrent to North Korean aggression. In large part, these systems are short-range tactical vehicles designed to enhance the effectiveness of South Korea's military forces, which are greatly outnumbered by the North Korean army. Autonomous and remotely piloted vehicles provide key assistance in securing the border and intelligence collection. The South Korean UAV program began with the purchase of Harpy models from Israel and a Shadow 400 system from the United States but has quickly grown to incorporate native procurement and production.

Robots offer substantial advantages over humans when placed in fixed locations. They are immune to fear, boredom, and corruption, three factors that play a major role in the DMZ. The Samsung Corporation's defense contracting division, Samsung Techwin, has developed the SGR-A1, a sentry robot armed with a light machine gun. Because the SGR-A1 is immobile, it is a far simpler machine than the wheeled and tracked systems utilized by many military forces around the world. However, given the fixed length of border to be guarded and the declining birthrate in South Korea, this may be the simplest and easiest solution to a decades-old problem. It can use infrared and digital cameras to sense potential targets and then notify its human operator of any anomalies. If ordered to engage, the robot's sensors allow attacks at more than two miles, a tremendously long distance for a 5.56-millimeter rifle. It can also be loaded with rubber bullets to provide a nonlethal response to incursions. Further refining its sentry functions, the robot has a microphone, speakers, and speech recognition capability, allowing it to challenge individuals for changeable passwords. Upon hearing an incorrect response, it can be programmed to sound an alarm or to automatically fire, potentially granting lethal decision making to the machine (Kumagai 2007; Page 2007).

The largest UAV developer in South Korea is the Uconsystem Company, based in Daejon. Uconsystem has produced both fixed-wing and rotary-wing designs. The RemoEye series offers a variety of small UAVs with extensive sensor packages. The largest of the series, the RemoEye 015, still weighs only 15 kilograms, with a range of 50 kilometers and an endurance of up to five hours. Uconsystem has also developed a crop-spraying helicopter, the RemoH-C100, which could quickly convert to the delivery of unconventional munitions in the event of a North Korean biological, chemical, or nuclear attack.

Larger UAVs have been developed by Korea Aerospace Industries (KAI) and Korea Aerospace Research Institute (KARI). The KAI Night Intruder 300 can carry almost 50 kilograms of sensory packages, including extremely advanced infrared sensors. It is currently fielded by both the army and the navy, both of which are concerned with preventing the infiltration of North Korean operatives across the border. The KARI Smart is still in development but is rare among UAVs because it utilizes tilt-rotors, allowing vertical takeoff and landing but much faster horizontal flight than traditional helicopter designs. Despite a wingspan of less than three meters, it is expected to carry a payload of nearly 100 kilograms.

Other Asian Countries

South Korea's erstwhile northern neighbor is one of the most secretive nations in the world. Long ostracized for its aggression, dictatorial oppression of its own populace, and pursuit of nuclear, chemical, and biological weapons, the North Korean government maintains its grip on power through military force. The North Korean army is one of the most disproportionately large forces in the world, with more than 1 million troops on active duty and a further 8 million in the reserves, out of a total population of only 24 million. It is likely that North Korea has pursued the development of UAVs and military robots, and the nation is in possession of Yakolev Pchelas purchased from Russia in 1995.

Thus far, virtually all research into UAVs in Malaysia has been carried out by Composites Technology Research Malaysia (CTRM). The Aludra, a multirole tactical UAV, was classified by the Malaysian government as "a national project with immediate effect" on October 21, 2008. This indicates the urgency the Malaysian Defense Ministry felt toward the development of remotely piloted vehicles. Its primary function is coastal monitoring, an absolutely vital role for a nation with Malaysia's topography. The Aviation Eagle ARV, also from CTRM, is a reconnaissance airplane that can be flown with or without a pilot. Like the Aludra, its primary purpose is coastal security (Daly 2009, 148).

Singapore's UAV fleet is largely, though not exclusively, produced by Singapore Technologies Aerospace. Most of its designs are small enough for hand launching and guidance by a single operator using a laptop computer. All serve primarily as ISR platforms, although the largest, the Skyblade IV, is also capable of search and rescue, coastal patrol, and artillery support missions. Cradance Services has designed the Golden Eagle, a micro-UAV weighing less than one kilogram but capable of loitering two hours and carrying either a camera, a gas analyzer, or a microphone, all designed to fit within its 80-gram payload capacity. Singapore has also imported a substantial number of UAVs from Israel, including the Blue Horizon, Hermes 450, Searcher II, and Skylark systems (Daly 2009, 184–187).

Perhaps in response to feared Chinese aggression, Taiwan has created a nascent research program into UAV technology at the Chang Shan Institute of Science and Technology (CSIST). Thus far, it has produced a small number of relatively light ISR aircraft, including the Chung Shyang and Kestrel series. Both are designed to carry an array of sensors for coastal patrol and have the ability to provide communication relays in the event that satellite systems are lost or jammed. Taiwan has also imported Searcher II UAVs from Israel (Daly 2009, 208–209; Minnick 2009).

Although Asian robotic and remotely piloted vehicles are not as numerous or as capable as those found in the United States and Europe, the region is quickly embracing the potential of these new technologies. China, in particular, is committed to developing a first-rate selection of autonomous weapons as a means to remain competitive in the global environment and to secure its own territory. In time, these devices may provide a power-projection capability currently lacking in the Chinese military, a fact that may provoke many of China's neighbors to accelerate their own development programs in response to a perceived threat.

South Pacific Region

Both Australia and New Zealand have begun to develop autonomous systems for military service. Given the vast distances and the relatively sparse population density of each nation, it is unsurprising that unmanned weapons systems and ISR platforms would be of immediate interest and utility for these nations. In Australia, a combination of government research facilities, private corporations, and academic institutions contribute to the development of robotic capabilities, particularly UAVs. New Zealand remains at the nascent stage, with a single corporation dominating the current development of unmanned systems.

Aerosonde has created one of the most durable and versatile UAVs on the global market and has seen a substantial number of orders from international customers, including several U.S. federal agencies, a number of American universities, and weather services around the world. Aerosonde's long-range, long-endurance platform can travel more than 2,900 kilometers and stay aloft for more than 30 hours, despite a system weight of only 15 kilograms. The system is even more attractive because it can be controlled using a pair of laptop computers, can be launched from the roof of virtually any automobile, and utilizes premium unleaded gasoline rather than ultrarefined aviation fuel or hard-to-acquire special materials. It can function with a wide variety of military or scientific sensors and is capable of autonomous takeoff, flight, and landings. The Australian military deployed Aerosondes to the Solomon Islands and East Timor in 2003.

Australia's need for a tactical UAV, demonstrated in coalition warfare in Afghanistan, has largely been filled by Codarra Advanced Systems, which has developed and marketed the Avatar, a low-cost ISR platform designed to be easily portable and quickly deployed. Because the unit has a battery-powered motor and is launched by hand, its logistical requirements are minimal. The UAV is controlled by a laptop computer through a Windows program, making it a far more user-friendly control system than most UAVs in current use. It is limited in speed (50 kilometers per hour), range (24 kilometers), and endurance (one hour), but it is capable of real-time imagery transmission. The Avatar saw duty in East Timor in 2003 (Australian Department of Industry 2007; Codarra 2010). Codarra has also developed the Silverback, an unmanned ground vehicle (UGV) designed to perform reconnaissance via remote control. Unlike many UGVs, the Silverback also has a manipulator arm, allowing it to deploy sensor devices in addition to its own internal video and audio sensors. It is capable of performing other functions, including target designation or direct weapons employment, depending on the package installed by the end user (Codarra 2011).

Another major developer in Australia is V-TOL Aerospace, creator of the i-Copter and Warrigal UAVs. The i-Copter is built in multiple sizes, from the 40-kilogram i-Copter Seeker, a dedicated ISR platform, to the i-Copter Phantom, a 250-kilogram helicopter that can perform ISR missions but can also engage in target acquisition and illumination as well as communication relay missions. The Warrigal is a battery-powered fixed-wing mini-UAV designed for short tactical missions. Its four-kilogram frame can carry only one kilogram of cargo, but the small propeller motor is almost silent, making this vehicle extremely difficult to detect, particularly because it has an all-weather capability that rivals any mini-UAV on the market (Daly 2009, 11–13).

Australia's Airborne Defense Research Organization (ADRO), later incorporated into Skytec Plastics, developed a short-range ISR and target acquisition system, the ADRO Pelican Observer, that is designed for rapid deployment and recovery. Although the aircraft is capable of a remarkably slow loiter speed for in-depth scouting and has a relatively long endurance, it also has an unwieldy logistical requirement of a full ground control station complete with four operators. As such, the Australian military has continued to seek further options to fulfill its ISR requirements.

In addition to developing its own systems, Australia has imported a number of UAV models, many of which have subsequently been deployed. Israel and the United States have been the primary suppliers of UAVs to the Australian military. In 2005, the Australian government purchased six Israeli Skylark systems, followed by an I-View 250 in 2008. The United States has sold RQ-11 Ravens and Insitu ScanEagles to Australia, which has deployed them to Iraq and Afghanistan.

New Zealand's TGR Helicorp, founded by a former member of the nation's parliament, Trevor Rogers, developed several versions of an autonomous helicopter that was designed for attack missions, ISR, and mountain rescues. These aircraft are noteworthy for their size; the Snark Mark III is capable of carrying a payload of 650 kilograms for more than 24 hours. Its operation range exceeds 3,700 kilometers. Perhaps more uniquely, the Snark's engine burns diesel fuel rather than aviation gasoline. However, despite the early promise of the Snark and Alpine Wasp UAVs, the corporation fell into receivership, and the aircraft have not been developed into full production. By 2008, the global market for UAVs was relatively crowded, and TGR Helicorp did not possess sufficient diversity or efficiency to remain competitive, despite early promise shown by test models (Daly 2009, 154–155; La Franchi 2009).

Middle East

While the use of ground robots by Middle Eastern nations is relatively rare, unmanned aerial vehicles have become extremely common in the region's skies. In particular, Israel has embraced the use of unmanned systems as a key mechanism to improve state security and compensate for its low population relative to its neighbors. Other early adopters of UAV technology include Bahrain, Iran, Jordan, Turkey, and the United Arab Emirates (UAE). Nonstate actors known to utilize UAVs include Hezbollah and Hamas, with the possibility that other nonstate entities have acquired such technology on the open market, regardless of efforts at weapons technology nonproliferation.

Israel

Israel maintains one of the most technology-dependent militaries in the world. Development of UAVs in Israel has largely been conducted by private firms. The corporations are spread throughout the nation, but most are relatively close to Tel Aviv or Haifa. The largest Israeli manufacturers of UAVs include Elbit Systems, Rafael Armament Development Authority, and Steadicopter in the north. Centrally located manufacturers include Israel Aerospace Industries, Israel Military Industries, EMIT Aviation Consult, Aeronautical Defense Systems, Blue Bird Aero Systems, Innocon, Top I Vision, and Urban Aeronautics. While the Israeli government works in close conjunction with each of these companies, development of unmanned weapons systems is not a state-run industry in Israel. The independent development model has created a wide variety of airframes and has in all likelihood allowed a level of creativity unmatched in the region; however, it can also prove unwieldy and undependable if corporate manufacturers choose to pursue other developments or cannot remain fiscally solvent.

Elbit Systems in Haifa has developed two major UAV lines for the Israel Defense Forces (IDF), the Hermes and the Skylark. The Hermes systems range from the Hermes 180, a short-range tactical UAV designed primarily for reconnaissance and weighing just over 200 kilograms, to the massive Hermes 1500, a 1,750-kilogram long-endurance platform capable of fully autonomous flight. In 2009, Hermes 450 UAVs were used to attack a convoy in Sudan carrying Iranian rockets to Hamas militants in Gaza (Mahnaimi 2009). The enormous Hermes 1500 can also function as a mobile communication hub, serving essentially the same function as a satellite relay system at a fraction of the cost and allowing disrupted or damaged communication lines to be quickly reestablished at any point within Israel's borders. While Israel has contracted to launch satellites in orbit, the Israeli government does not have the capability to conduct independent launches, and as such, remains at the whims of other states in the event that its space-based capabilities become degraded or destroyed. The Skylark I and II are both lightweight tactical scout craft, characterized by extremely advanced sensory capability. Both have proven successful in military operations and have prompted orders from several militaries around the globe (Daly 2009, 96–104).

The Steadicopter Black Eagle 50, also constructed near Haifa, is a small rotary-wing ISR platform capable of hovering in a single location for nearly three hours. It is capable of fully autonomous flight but does have a limited range barely exceeding 10 kilometers and a payload capacity of only five kilograms. This small unit combines the unobtrusive positioning of an aerostat with the maneuverability of a helicopter. While its sensor package is necessarily limited, and its range prohibits long patrol routes, it can serve an invaluable function as an overwatch platform, or it can be deployed at relatively slow speeds to scout in advance of ground units. Rafael Armament Development Authority, also in Haifa, has created the Rafael SkyLite, a catapult-launched mini-UAV that can be transported in two backpacks and guided by an unmodified laptop computer. While its endurance seems short at only 90 minutes, in reality, the loiter time is long relative to the size of the vehicle. It can be recovered and relaunched in a matter of only a few minutes, due in large part to the use of an electric motor powered by lithium-polymer batteries (Rafael 2005). UAVs of this type are in constant demand by land units, as they can provide almost immediate intelligence about the surrounding environs but do not require substantial logistical support. As of 2010, Rafael decided to abandon its UAV development programs and concentrate on producing payloads for other companies in the UAV market (Daly 2009, 120–121; Egozi 2010).

Israel Aerospace Industries (IAI), the largest UAV developer in the Tel Aviv region, also offers the most diverse UAV product line of any Israeli defense contractor. Its smallest machine, the IAI Mosquito, is a micro-UAV weighing only half a kilogram. Despite its size, this tiny hand-launched aircraft carries a video camera with a real-time data link, providing immediate surveillance to be transmitted to tactical commanders and essentially allowing a peek around corners or over terrain features without placing troops at risk. IAI also offers some of the largest UAVs in the world, such as the IAI Heron TP Eitan, a multirole UAV weighing nearly 5,000 kilograms. This behemoth can function as a long-endurance ISR system, given its ability to remain aloft for 36 hours. However, it is also capable of carrying direct attack munitions in much the same manner as the better-known U.S. MQ-1 Predator and MQ-9 Reaper systems. The TP Eitan can carry ballistic missile detection and interception equipment, which are absolutely vital to the IDF in light of the development of long-range ballistic missiles in Iran and the use of medium- and short-range ballistic missiles against Israeli targets by Hezbollah units in Lebanon. This aircraft can also serve in an aerial tanker capacity, providing aerial refueling for both manned and unmanned systems. With a wingspan of over 26 meters, the ability to operate at altitudes of up to 14,000 meters, and a payload capacity of 1,000 kilograms, this fully autonomous UAV is one of the most complex systems currently in operation anywhere in the world. It made its operational debut in 2009, providing reconnaissance during the Israeli antirocket campaign in Gaza (Daly 2009, 106–116; Opall-Rome 2010).

In addition to traditional fixed-wing UAVs, some Israeli developers have begun to experiment with more radical designs, some of which draw on much earlier forms of manned flight. Top I Vision has produced the Rufus, a helium-filled aerostat designed for static surveillance and artillery fire correction. Because the airship has no rigid frame, it can be stored in a box of half a cubic meter and can be quickly readied for deployment. It has virtually no radar signature, due to the materials used in its construction, and is thus difficult to destroy from the ground. Thus far, it has proven most useful in traffic-control and crime monitoring situations, although it certainly offers military appeal

as well. Given the topography of Israel's borders, simple surveillance aerostats operating at sufficiently high altitudes offer a
low-cost solution to the problem of monitoring wide expanses of
sparsely settled ground (Daly 2009, 122–125).

The UrbanAero AirMule, built by Urban Aeronautics, is a
VTOL aircraft that can offer aerial resupply and medical evacu-
ation to troops in combat. It can carry more than 200 kilograms
of supplies, allowing the rapid withdrawal of two battlefield
casualties as soon as they can be stabilized for transport. It can
be remotely flown but is also capable of preprogrammed flight,
landing safely by itself when given a beacon to home in on.
Because it is an unmanned system, it is much smaller, with the
corresponding savings in operational requirements, than a helicopter or fixed-wing aircraft with the same payload capacity, thus
providing a great deal more flexibility in where it can take off or
land. Perhaps the most salient feature of the aircraft is its use of
internal lift rotors—these make the aircraft able to operate in tight
spaces that would be inaccessible to traditional rotorcraft, as the
rotors are protected from striking objects. The design also reduces
engine noise and offers a significantly reduced radar profile, making it ideal for the type of urban operations the Israelis have faced
in Gaza and the West Bank (UrbanAero 2011).

Jordan

UAV development in Jordan is largely controlled by Jordan Aerospace Industries (JAI). Thus far, JAI has developed two systems,
the Silent Eye and the Jordan Falcon, both actually manufactured
by Jordan Advanced Remote Systems (JARS) in Amman. The
Silent Eye, a mini-UAV, has a short endurance and range but is
easily transportable and requires only two ground crew members. The Jordan Falcon can carry ISR packages but is also an
effective mobile communication jammer or communication relay
system, as needed. While it is normally piloted remotely, it is
capable of autonomous flight and has an extremely long range
of nearly 500 kilometers. However, while the system is capable
of maintaining a real-time data link, the Jordanian government
does not currently have the satellite transmission capability necessary to extend the data link beyond line of sight, which limits
the utility of the real-time connection to a range of approximately
50 kilometers (Daly 2009, 138–140).

Turkey

Like Jordan, Turkey has a single primary UAV developer, Turkish Aerospace Industries (TAI). The Turkish military has focused on developing a small number of multirole aircraft rather than a host of specialized vehicles. While this substantially reduces overall development costs, it also tends to reduce the effectiveness of the UAV within individual missions. Turkey does gain some technical development assistance through its position as a member of the North Atlantic Treaty Organization (NATO), the only such member within the Middle East region. The TAI Gözcü is a short-range tactical system that requires two ground operators, one to fly the aircraft and the other to control the sensors. It requires an extensive ground control station, including a truck-mounted launcher and flight trailer, which creates a substantial logistical hurdle for a moderate ISR capability. While further developments will likely reduce the ground control requirements for this aircraft, there does not appear to be substantial impetus for developing more sophisticated models in a rapid time frame, as the Turkish government remains committed to manned military systems. The TAI ANKA, a much larger and more flexible aircraft, closely resembles the U.S. RQ-1 Predator. Not only can it carry a wide variety of sensor packages, it can also carry a laser target designator or direct attack munitions. With a payload capacity of 200 kilograms, an endurance capability of 24 hours, and a powerful long-range data link, the ANKA is a much more formidable UAV within its class than the Gözcü, suggesting that the Turkish government may prefer to focus on large systems for most future development efforts. Turkey has purchased Gnat, I-Gnat, and Shadow 600 UAVs from the United States as well as Herons from Israel to augment its nascent UAV capabilities (Daly 2009, 210–211; TAI 2011).

Iran

The Iranian weapons industry is second in the region only to the Israeli munitions system. However, military research is much more tightly regulated in Iran, meaning most development has been performed by the Iran Aircraft Manufacturing Industries (HESA) or its subsidiary, Qods Aviation Industries (QAI). This development model has allowed a greater amount of secrecy for developments but also restricts the development of radical designs and competing visions of autonomous vehicles. The HESA Ababils are multirole UAVs that can perform both ISR and direct attack missions. While they have a relatively short endurance, the Ababil systems are characterized by extremely rapid flight, particularly UAVs, with the Ababil-II crusing at 370 kilometers per hour. Iran has provided a small number of these craft to Hezbollah for use against Israeli targets. In 2009, an Ababil-3 crossed into Iraqi airspace and was eventually shot down by an American warplane (Hodge 2009). Iran has also developed the Qods Mohajer systems, which are much smaller ISR platforms primarily designed to monitor Iranian borders. They carry light sensor payloads and have fairly limited ranges, with the largest version, the Mohajer 4, capable of a radius of almost 160 kilometers for a maximum of seven hours.

UAE

Much like Israel, the United Arab Emirates (UAE) has turned to technological solutions as a means to overcome a small population and a difficult geographic position. While the UAE research and development systems are not as robust as those of their Israeli competitors, the nation is making rapid strides in sophistication and has begun to market its UAVs to potential foreign purchasers. With the UAE's vast mineral wealth and lack of a serious threat to state security, the decision to develop and sell UAVs may be primarily an attempt to diversify the UAE's economy and begin building a stronger high-technology sector. The UAE began its UAV programs by importing Seekers from Israel, APID-55s from Sweden, and CamCopter S-100s from Austria. It deployed the CamCopters to Afghanistan in support of its small ground forces there in 2006. It has also begun to develop a domestic production capability.

Advanced Target Systems (ATS) is the primary developer and manufacturer of UAVs for the UAE. The company, based in Mussafah, Abu Dhabi, has largely focused on designs that allow substantial mission endurance. The smallest of their airframes, the ATS Yabhon-H, has a payload capacity of only five kilograms, yet it can fly for eight hours of tactical surveillance. At the other end of the spectrum, the ATS Yabhon-RX can carry 60 kilograms on a 320-kilogram airframe and is designed to perform ISR, search and rescue, and environmental monitoring missions. It requires only 150 meters of runway for launch and can fly for nearly two full days on a single mission (Daly 2009, 217–219).

North America

Although the United States is by far the largest developer, producer, operator, and exporter of unmanned military vehicles, it is not the only nation in North America that is vigorously pursuing such technology. Both Canada and Mexico have developed a number of models, including some systems that are now being exported on the global market. Thus far, Mexican unmanned systems serve an ISR function and have been used in dual military and law enforcement roles. They have proven useful in attempts to stymie the international drug trade, both for surveillance and in a tactical role supporting Mexican military forces engaged in combat with narcotics cartels. American border patrol units have also begun to deploy unmanned systems to the Mexican border, in part to slow immigration but largely in response to gang-related violence in northern Mexico that has spilled across the U.S. border in recent years.

Canadian UAV efforts have developed some sophisticated unmanned cargo systems. The MMIST Sherpa, developed and manufactured by Mist Mobility Integrated Systems Technology, is a glider drone with a postrelease range of 20 kilometers, an accuracy of under 100 meters error, and a payload capacity of 1,000 kilograms. Essentially, this aircraft allows the aerial resupply of ground forces through traditional airdrop methods but does not require the manned dropping systems to enter contested airspace. Further, the trajectory of the glider makes it far less vulnerable to ground fire than traditional pallet and parachute cargo airdrops. The system's versatility and heavy lift capability have attracted the interest of the U.S. military, as well as that of other North Atlantic Treaty Organization nations. U.S. forces have deployed Sherpas in Iraq and reported tremendous success using the systems for rapid-response aerial resupply to fielded units (Daly 2009, 24–25).

The same Canadian corporation created the SnowGoose, a massive self-propelled, unmanned cargo transport that mounts six external bins, each of which can hold nearly 50 kilograms of supplies. The external mount location allows extremely rapid loading and unloading of the aircraft, and the bins can also be fitted for aerial release and parachute descent if landing the aircraft is impossible or undesirable. Like the Sherpa, the SnowGoose has attracted American military interest, specifically the U.S. Special Operations Command, which has agreed to purchase up to 200 vehicles. The vast distances in Canada have also triggered the development of micro-UAVs for crop and forest monitoring, with the MicroPilot CropCam the most commercially successful model. Currently, the CropCam has been exported to at least 18 nations (Daly 2009, 21–23).

In addition to its own domestic developments, the Canadian military has also utilized a number of imported models. The Canadian army has deployed Israeli Skylarks and French Sperwers to Afghanistan in support of the Global War on Terror. The Canadian air force also has purchased Altair and Silver Fox UAVs from American suppliers.

South America

South America does not have a substantial number of nations aggressively pursuing the use of robotic and drone systems in their military forces. In part, this is due to the relatively small military budgets in most South American nations and the general level of international security on the continent. The region is insulated geographically from most global conflict zones, and current conflicts in the area are primarily internal insurgencies that rarely escalate into conventional conflict. A limited amount of cross-border fighting has occurred in the past few decades, largely as outgrowths of insurgencies or the activities of narcoterrorist organizations. Because most South American nations have relied on relatively low-technology solutions to their counterinsurgency problems, robots and drones have not significantly altered the military forces of the region. However, growing international tensions fueled by human smuggling and illicit narcotics transportation may provoke some South American militaries to reexamine the potential benefits of relatively low-cost UAV systems that might assist in the monitoring of border crossings or relatively unsettled regions in challenging terrain.

The primary South American exceptions to the lack of unmanned systems are Brazil and Argentina, the two most militarily advanced nations in the region. Brazilian drone systems include short-range tactical observation platforms, such as the Watchdog and AvantVision ISR UAVs. A more recent development in the Brazilian arsenal is the FS-03 Starcopter, a helicopter UAV weighing nearly 150 kilograms and capable of lifting payloads in excess of 100 kilograms. The model, first produced in 2007, remains flexible enough to serve in a wide variety of roles and is capable of autonomous flight. Argentinian models, although smaller than Brazilian products, include a more diverse array of vehicles. At least two mini-UAV models, both developed by Nostromo Defensa, have been exported for foreign demonstrations. Argentina is currently the only South American nation exporting UAV models. Chile, Ecuador, and Venezuela are all developing UAV technology, and the first Chilean UAV systems have entered the manufacturing stage of production.

Africa

African militaries have not adopted unmanned weapons systems in large numbers to date. While several African nations have small development programs in place, and a few have imported models for testing purposes, only the Republic of South Africa (RSA) has developed its systems enough to export them on the world market. Egypt and Tunisia have begun to manufacture relatively unsophisticated ISR platforms. Algeria and Botswana have formally incorporated UAVs into their military forces. Brunei, Libya, and Morocco have begun to examine the potential applications of UAVs, but only Libya has created a true development program aimed at production of operational models in the near future. The 2011 unrest in Libya seems to have virtually eliminated the nascent development program, although loyalist Libyan forces flew systems imported from Pakistan during the conflict. Coalition members enforcing a no-fly zone over Libya utilized UAVs, both for ISR purposes and in ground-attack roles. Rebel military elements also utilized tactical UAVs, including a Canadian-built Aeryon Scout micro-UAV (DefenceWeb 2011).

Long the most technologically advanced African military, the armed forces of RSA have demonstrated the most interest in

UAVs. Two South African corporations, Advanced Technologies & Engineering (ATE) and Denel, have thus far dominated the manufacturing of new systems. ATE has created a host of tactical and ISR systems, including the ATE Vulture, which has been adopted by the South African army. The Vulture, a 125-kilogram ISR platform, can be integrated with G5 and G6 howitzers, allowing it to serve as an unmanned forward artillery observer, illuminating targets and offering corrections. The ATE Sentinel series includes a variety of airframe sizes and payload capacities, and includes a modular design that allows a rapid change of sensor equipment, making the craft true multirole UAVs (Daly 2009, 189–193; Space War 2006). The Denel Bateleur is the largest UAV produced in Africa. It has a 15-meter wingspan and a maximum takeoff weight of 1,000 kilograms, including 200 kilograms of payload. This multirole UAV can cruise for up to 24 hours, allowing sustained ISR operations over open water. The smaller Denel Seeker also has a long endurance but a much shorter range due to a weaker ground control system (Daly 2009, 193–196; Engelbrecht 2010).

Most African nations do not yet have the capability to produce UAVs, but that has not stopped some from acquiring and using them in their military forces. The Algerian Air Force flies both the Seeker I and II models, imported from the RSA. Botswana has purchased the Hermes 450 from Israel. The Egyptian army flies primarily imported UAVs, although it has begun to domestically produce target drones. It has acquired slightly modified Teledyne Ryan Fire Flies (renamed Scarabs) from the United States; Sky Eyes from the United Kingdom; and CamCopters from Austria. The Nigerian coast guard utilizes Aerostars that it acquired from Israel in 2006.

The vast distances and large amount of conflict in Africa may lead to increasing efforts to develop UAVs domestically in African nations. Also, as the technology continues to grow and increase in capabilities, the world UAV market is likely to become inundated with models considered old or even obsolete by first-rate militaries. In much the same fashion that many African nations have found other forms of cast-off military technology to be sufficient for their defense requirements, it is almost certain that they will be able to obtain UAVs relatively soon. These UAVs may serve a useful function patrolling borders and monitoring population migrations and will likely prove attractive to nations bordering the most conflict-ridden regions, hoping to avoid a spillover onto their own soil.

Conclusion

The proliferation of robotic and drone technology around the globe has largely followed the pattern established by earlier military advances. There is nothing revolutionary about the expansion of these vehicles into the military forces of even the smallest nations on earth. While the development of these devices does not require the level of investment of technological skill as do unconventional weapons such as nuclear, biological, and chemical weapons, neither do they inspire the international condemnation that typically accompanies the pursuit of weapons of mass destruction. Likewise, the possession of unmanned military vehicles does not convey such a significant advantage that they are absolutely necessary for conflict in the modern world. As is true of manned vehicles and other types of weaponry, nations with the largest military forces, the most diverse manufacturing bases, and the most high-technology industry are the most likely to pursue these devices.

Because robots and drones come in such a wide variety of designs and capabilities, they have become readily available to any nation or even a well-funded nonstate actor that chooses to purchase them on the international market. By far, UAVs have become the most common military robots around the world. Domestic development of basic UAV systems is relatively easy, as they can be manufactured using off-the-shelf components to create at least a rudimentary capability. The proliferation of UAVs demonstrates that while this type of technology is constantly improving, it has also matured to a level that unmanned ground and undersea vehicles have yet to reach. Military robots of these types remain rare and have not gone into substantial production in most regions of the globe, though they are likely to do so in the near future.

Despite the continual proliferation of autonomous weaponry, the United States remains far and away the largest manufacturer of these systems, and the American designs are the most capable in the world. For the foreseeable future, the United States is likely to remain a dominant user of autonomous military vehicles, although there are certainly major competitors on the international arms market, particularly within specific niches of RPV production. The military forces of the world, like the U.S. military, have been quick to recognize the potential advantages of unmanned systems but have not abandoned their traditional force structures to incorporate the new machines. Rather, they have integrated them into existing organizations, adding capabilities without triggering a robotics revolution.

References

AirSerbia. 2005. *UAV: IBL-2004 Serbian Made Unmanned Aerial Vehicle at Partner 2005*. Retrieved from http://airserbia.com/video/display/35/news.

ALCORE Technologies. 2011. *ALCORE UAVs*. Retrieved from http://www.alcoretech.com/.

AleniaAeronautica. 2008. *Sky-X*. Retrieved from http://www.alenia-aeronautica.it/eng/Media/Scheda%20tecnica/sky-x%20data%20sheet.pdf.

Australian Department of Industry. 2007. *Unmanned Technologies Australia: Industry Capabilities*. Retrieved from http://www.uavm.com/images/capability_brochure__20050601153547.pdf.

Aviation Week. 2010. *India Eyes Armed Rustom UAV* (October 25). Retrieved from http://www.aviationweek.com/aw/generic/story.jsp?topicName=unmanned&id=news/awst/2010/10/25/AW_10_25_2010_p30-263965.xml&headline=India%20Eyes%20Armed%20Rustom%20UAV&channel=&from=topicalreports.

BAE Systems. 2011. TARANIS: Informing the Future Force Mix. Retrieved from http://www.baesystems.com/BAEProd/groups/public/documents/bae_publication/bae_pdf_taranis_fact_sheet.pdf.

Codarra. 2010. *Avatar Unmanned Aerial Vehicle*. Retrieved from http://www.codarra.com/index.php?option=com_content&view=article&id=271&Itemid=366.

Codarra. 2011. *Silverback Unmanned Ground Vehicle*. Retrieved from http://www.codarra.com/index.php?option=com_content&view=article& id=272&Itemid=367.

CybAero. 2011. *The APID 55 VTOL UAV System*. Retrieved from http://www.cybaero.se/english/dokument/produktblad/ProductsheetAPID55.pdf

Daly, Mark, ed. 2009. *Jane's Unmanned Aerial Vehicles and Targets*. Alexandria, VA: Jane's Information Group.

Dassault Aviation. 2006. *nEUROn: The Program Goal*. Retrieved from http://www.dassault-aviation.com/en/defense/neuron/the-programm-goal.html?L=1.

Davenport, Justin. 2007. "Tens of Thousands of CCTV Cameras, Yet 80% of Crime Unsolved." *Evening Standard Online* (September 19). Retrieved from http://www.thisislondon.co.uk/news/article-23412867-tens-of-thousands-of-cctv-cameras-yet-80-of-crime-unsolved.do.

DefenceWeb. 2011. "Libyan Rebels Using Micro UAV." *DefenceWeb* (August 25). Retrieved from http://www.defenceweb.co.za/index.php?option=com_content&view=article&id=18339&catid=35&Itemid=107.

Diamond, Jared. 1998. *Guns, Germs, and Steel: The Fate of Human Societies*. New York: W. W. Norton.

EADS (European Aeronautic Defence and Space Company). 2003. *EADS Systems & Defence Electronics at the Le Bourget International Air Show 2003 in Paris, France*. Retrieved from http://www.eads.com/eads/int/en/news/press.21014e83-bdd6-4270-9a81-8ff6f4fa52b0.08af92a7-2c53-400a-8429-8b135733cbcc.html?queryStr=orka-1200&pid=1.

EADS (European Aeronautic Defence and Space Company). 2009. *EADS Defence & Security Tests the Largest Unmanned Aerial System (UAV) Ever Built in Europe*. Retrieved from http://www.eads.com/eads/int/en/news/press.41155926-4607-48a1-ac5e-9b44d3a94770.08af92a7-2c53-400a-8429-8b135733cbcc.html?queryStr=barracuda&pid=1.

Egozi, Arie. 2010. "Rafael Quits Israel's Congested UAV Market." *Flight International* (January 19). Retrieved from http://www.flightglobal.com/articles/2010/01/19/337276/rafael-quits-israels-congested-uav-market.html.

EMT Penzberg. 2011. *LUNA UAV System*. http://www.emt-penzberg.de/index.php?14&L=1.

Engelbrecht, Leon. 2010. "Denel Unveils Armed Seeker, Impi Missile." *DefenceWeb* (September 24). Retrieved from http://www.defenceweb.co.za/index.php?option=com_content&view=article&id=9750:denel-unveils-armed-seeker-impi-missile&catid=35:Aerospace&Itemid=107.

FanWing. 2011. *Prototype UAV Airframes*. Retrieved from http://www.fanwing.com/uav.htm.

FlightGlobal. 2004. *Unmanned Systems: Europe 2004: German Army Poised to Reorganise UAV Forces* (May 18) Retrieved from http://www.flightglobal.com/channels/mro/articles/2004/05/18/182110/unmanned-systems-europe-2004-german-army-poised-to-reorganise-uav-forces.html.

FlightGlobal. 2006. *Aladin UAVs to Support Netherlands Troops in Afghanistan* (February 5). Retrieved from http://www.flightglobal.com/articles/2006/05/02/206283/aladin-uavs-to-support-netherlands-troops-in-afghanistan.html.

FlightGlobal. 2011. *UAV Directory: Aircraft Specification: Cyberflight S.O.D. IV.* Retrieved from http://www.flightglobal.com/directory.

Genuth, Iddo. 2007. "Camcopter S-100 UAV." *The Future of Things* (August 31). Retrieved from http://thefutureofthings.com/pod/288/camcopter-s-100-uav.html.

Hanlon, Mike. 2004. "Yamaha's RMAX: The World's Most Advanced Non-Military UAV." *Gizmag*. Retrieved from http://www.gizmag.com/go/2440/.

Hodge, Nathan. 2009. "U.S. Military Confirms It Shot Down Iranian Drone." *Wired* (March 16). Retrieved from http://www.wired.com/dangerroom/2009/03/confirmed-us-do/.

Hutchison, Harold C. 2005. "Ambitious Russian UAV Programs." *Strategy Page* (November 26). Retrieved from http://www.strategypage.com/htmw/htairfo/articles/20051126.aspx.

IAS (International Aviation Supply). 2011. Retrieved from http://www.interaviosup.it/.

Indian Military. 2010. "AK-630 Gatling Gun Close in Weapon System." *Indian Military Online* (March 12). Retrieved from http://www.indian-military.org/navy/ordnance/close-in-weapon-system/267-ak-630-gatling-gun-close-in-weapon-system.html.

Integrated Dynamics. 2010. *Surveillance UAV Systems.* Retrieved from http://www.idaerospace.com/suav.html.

Irkut Corporation. 2010. *Unmanned Aerial System Irkut-850.* Retrieved from http://www.irkut.com/common//img/uploaded/files/IRKUT-850_2010.02_eng.pdf.

Kenyon, Henry S. 2005. "Sojka Spreads Its Wings." *Signal Online* (September). Retrieved from http://www.afcea.org/signal/articles/templates/SIGNAL_Article_Template.asp?articleid=1020&zoneid=124.

Kulkarni, Prasad. 2008. "Daksh Could Be Useful in Mumbai Operations." *Times of India Online* (November 28). Retrieved from http://articles.timesofindia.indiatimes.com/2008-11-28/pune/27893859_1_vehicle-daksh-detection-and-disposal.

Kumagai, Jean. 2007. "A Robotic Sentry for Korea's Demilitarized Zone." *IEEE Spectrum* (March). Retrieved from http://spectrum.ieee.org/robotics/military-robots/a-robotic-sentry-for-koreas-demilitarized-zone.

La Franchi, Peter. 2009. "Snark Seeking a New Home." *Shephard Group Online* (January 9). Retrieved from http://www.shephard.co.uk/news/1452/.

Mahnaimi, Uzi. 2009. "Israeli Drones Destroy Rocket-Smuggling Convoys in Sudan." *Times* (London) (March 29). Retrieved from http://www.timesonline.co.uk/tol/news/world/africa/article5993093.ece.

Martin, Matt J. 2010. *Predator: The Remote-Control Air War Over Iraq and Afghanistan: A Pilot's Story.* Minneapolis, MN: Zenith Press.

McNeill, William H. 1982. *The Pursuit of Power: Technology, Armed Force, and Society since A. D. 1000.* Chicago: University of Chicago Press.

Minnick, Wendell. 2009. "Taiwan's CSIST Shows Off Missiles, UAVs at TADTE." *Defense News* (August 16). Retrieved from http://www.defensenews.com/story.php?i=4237571.

Opall-Rome, Barbara. 2010. "Israel's Heavy-Hauling UAVs Are Ready for Battle." *Defense News* (January 25). Retrieved from http://www.defensenews.com/story.php?i=4469090&c=MID&s=AIR.

Page, Lewis. 2007. "South Korea to Field Gun-Cam Robots on DMZ." *Register* (London) (March 14). Retrieved from http://www.theregister.co.uk/2007/03/14/south_korean_gun_bots/.

Palmer, Brian. 2010. "Big Apple Is Watching You." *Slate* (May 3). Retrieved from http://www.slate.com/id/2252729.

Parker, Geoffrey. 1988. *The Military Revolution: Military Innovation and the Rise of the West, 1500–1800.* New York: Cambridge University Press.

Pyadushkin, Maxim. 2010. "Russian Military Plans MALE UAV Development." *Aviation Week Online* (January 26). Retrieved from http://www.aviationweek.com/aw/generic/story.jsp?id=news/awst/2010/01/25/AW_01_25_2010_p31-.

QinetiQ North America. 2011. *Unmanned Aerial Vehicle (UAV): The Zephyr.* Retrieved from http://www.qinetiq-na.com/products-zephyr.htm

Rafael (Rafael Advanced Defense Systems). 2005. *Rafael Presents Skylite B* (November 21). Retrieved from http://www.rafael.co.il/Marketing/192-1230-en/Marketing.aspx.

Roggio, Bill and Alexander Mayer. 2009. "Analysis: A Look at US Airstrikes in Pakistan through September 2009." *Long War Journal* (October 1). Retrieved from http://www.longwarjournal.org/archives/2009/10/analysis_us_airstrik.php.

RUAG (RUAG Aerospace Defence Technology). 2006. *Ranger UAV System.* Retrieved from http://www.ruag.com/de/Aviation/Military _Aviation_CH/Platforms/Unbemannte_Luftfahrzeuge/P_Ranger _UAV_System.pdf.

Saab. 2010. *Skeldar V-200 Land*. Retrieved from http://www.saabgroup.com/en/Air/Airborne-Solutions/Unmanned_Aerial_Systems/Skeldar_V-200_Land/.

Sagem (Sagem Défense Sécurité). 2011. *"Tactical UAV Systems*. Retrieved from http://www.sagem-ds.com/spip.php?rubrique90&lang=en.

SELEX Galileo. 2011. Retrieved from http://www.selex-sas.com/SelexGalileo/EN/Corporate/About_Us/index.sdo.

Singer, P. W. 2009. *Wired for War: The Robotics Revolution and Conflict in the Twenty-First Century*. New York: Penguin Books.

SIPRI (Stockholm International Peace Research Institute). 2011. Retrieved from http://armstrade.sipri.org/armstrade/page/toplist.php.

Space War. 2006. "South African Vulture UAV for Production." *Space War Online* (August 13). Retrieved from http://www.spacewar.com/reports/South_African_Vulture_UAV_for_Production_999.html.

Strategy Page. 2008. *Soaring Dragon Chases Global Hawk* (October 21). Retrieved from http://www.strategypage.com/htmw/htairfo/articles/20081021.aspx.

TAI (Turkish Aerospace Industries). 2011. *Unmanned Aerial Vehicle System Anka* (May). Retrieved from http://www.tai.com.tr/ds_Resource/image/content/ANKA-ENG.pdf.

Tasuma UK. 2009. *Surveillance Aircraft: Hawkeye*. Retrieved from http://www.tasuma-uk.com/tasuma.php?p=32.

Tupolev PSC. 2011. *On Progress of Unmanned Aircraft in Tupolev Design Bureau*. Retrieved from http://www.tupolev.ru/English/Show.asp?SectionID=66.

UrbanAero. 2011. *AirMule*. Retrieved from http://www.urbanaero.com/category/airmule.

Von Kospoth, Nicholas. 2009. "Chinese Researchers Break through the Mysteries of UAVs and UCAVs." *Defence Professionals Online* (October 14). Retrieved from http://www.defpro.com/daily/details/424/.

Warfare.ru. 2007. *MiG SKAT Unmanned Aerial Fighting Aircraft* (August 23). Retrieved from http://warfare.ru/?catid=324&linkid=2562

Warrior Aero-Marine. 2010. *Gull UAV*. Retrieved from http://www.warrioraero.com/GULL/index.htm.

ZALA Aero. 2009. *A-Level Aerosystems*. Retrieved from http://www.zala.aero/media/zalaaero-eng.pdf.

4

Chronology

800–700
BCE
Homer's *The Iliad* includes passages describing robot-like automatons. The fighting force of Myrmidons, led by Achilles, are likened to automatons in some translations.

700–600
BCE
Automatic theaters utilizing water-powered machines for special effects are constructed in Mesopotamia. Theaters demonstrated the potential of mechanical automata, far in advance of modern expectations.

500–400
BCE
Sophisticated water clocks utilizing feedback control systems are developed and put into use. While the earliest water clocks involved a simple outflow system, and measured time by the flow of liquid out of a pierced container, this new concept allowed the use of gears and dials to display the time, while accounting for changing lengths of days throughout the year. Later innovations allowed alarms using gongs, trumpets, or other noise-makers. These timepieces were much more sophisticated and dependable than the use of a sand glass.

ca. 400 BCE
Archytus designs an automaton shaped like a wooden dove, utilizing compressed air to power its movements. The shift from water to compressed air as a power source is a substantial technological leap and required a great deal more precision in the design and construction of automatic devices.

300–200 BCE	Archimedes of Syracuse develops the Archimedean Screw, a worm gear capable of bringing groundwater to the surface, and begins the study of hydrostatics. The ready supply of water, even in the absence of rivers or lakes, made reliance on hydraulic machines much more plausible.
210–209 BCE	An army of terracotta sculptures is buried in the tomb of Emperor Qin Shi Huang of China to help guard him in the afterlife. Over 8,000 of the figures have been discovered, representing a desire to replace human warriors with machines.
ca. 30–70 CE	Hero(n) of Alexandria constructs a rudimentary rocket engine (the aeolipile), a primitive steam engine, and a wind-powered organ. While serving no practical purpose, the development of these innovations had the potential to spur massive technological advances, but such study was interrupted by the decline of the Roman Empire.
ca. 850	The Banu Musa produce *The Book of Ingenious Devices* in Baghdad, working on behalf of the Abassid Caliph. The work consolidated Hellenistic texts on mechanical devices and also included many devices invented by the Banu Musa brothers. The book demonstrates the shift of technological innovation to the Middle East, where educational advances continued apace for centuries while Europe struggled to maintain literacy in even a fragment of the populace.
1206	Abu al-Jazari publishes *The Book of Knowledge of Ingenious Mechanical Devices* in Mesopotamia. The work includes detailed instructions for more than 50 machines, including automata. Unlike previous automatons, al-Jazari's inventions were designed for practical application, rather than simple amusement. As such, this work represents the maturation of ideas initially proposed and developed centuries earlier.

1497   Leonardo da Vinci designs an automaton knight, powered by springs and pulleys. His device demonstrated the return of technological innovation to Europe during the Renaissance, which drove technological development for centuries.

1515   Leonardo da Vinci builds a lion automaton for presentation to King Francis I of France. Patronage of European nobility provided the funds necessary for such creations and remained common practice for independent inventors for centuries.

1642   Blaise Pascal designs a rudimentary calculator capable of addition and subtraction. The mechanization of arithmetic, and later higher functions, greatly spurred scientific and engineering innovations.

1672   Gottfried Wilhelm von Leibniz develops the Stepped Reckoner, a calculator capable of addition, subtraction, multiplication, and division. This was a major improvement to Pascal's design and came only three decades after the earlier calculator's debut. The more complex relationships of multiplication and division required a far more complicated device.

1737   Jacques de Vaucanson builds The Flute Player, The Tambourine Player, and The Digesting Duck, three automata, for the amusement of the gentry in Paris. The disturbing realism of the devices both frightened and fascinated Parisian society. To increase the effect, de Vaucanson covered the musicians in real skin. The duck was later proven to contain a separate pouch of "digested food" for defecation; the corn placed in its mouth remained unchanged in the stomach compartment.

ca. 1760   Friederich von Knauss creates an automaton capable of writing short phrases as well as a mechanical musician. The ever-increasing demand for these types of amusements provided the funding necessary to allow advances in sophistication, which is reflected

ca. 1760   by the short period between de Vaucanson's devices
(*cont.*)   and those of Knauss.

1768–1774   Swiss watchmaker Pierre Jaquet-Droz creates
             three androids, The Writer, The Musician, and The
             Draughtsman. They incorporated a programmable
             memory system, arguably creating the first rudimen-
             tary computer. Because they allowed programming,
             these could also be considered the first robots in
             history, although they did not truly interact with their
             environment in any meaningful fashion.

ca. 1800   Henri Maillardet creates automatons capable of writ-
             ing and drawing, powered by springs and using a
             cam-based memory system. These creations still did
             not interact with their environment, despite appear-
             ances. In many ways, they were less sophisticated
             than a mechanical loom, but their resemblance to
             human figures caused a stir throughout Europe.

             Alessandro Volta develops the first electrochemical
             cell, the forerunner to the modern electrical battery,
             as a means to store electricity.

1805   Joseph Marie Charles Jacquard invents the
             punch card programming system for mechanical
             looms. The concept was later adapted to program-
             ming other machines. The complex patterns
             needed for woven fabrics could be created in advance
             and then replicated without error, very quickly.
             The mechanical loom thus performed a dangerous,
             dull, and dirty task, much like later robots and
             drones.

1822   Charles Babbage designs the Difference Engine, a
             massive calculator designed to mechanically derive
             numerical tables. Less than two centuries after the
             first rudimentary calculator, Babbage's machine
             could theoretically perform enormously complex
             mathematical functions without error, including
             using polynomials to approximate logarithmic and

trigonometric functions. Due to the costs involved, a working difference engine was not completed until 1991, when the London Science Museum built a working model to commemorate the 200th anniversary of Babbage's birth.

1825　William Sturgeon develops the first electromagnet, making possible the invention of the electric motor. This, in turn, led to propulsion systems for robots and unmanned vehicles.

1837　Charles Babbage first describes his Analytical Engine, the forerunner to modern general-purpose computers. Like the Difference Engine, it was never constructed due to funding limitations. There has been substantial debate over whether the machine would function as designed. In 2010, John Graham-Cumming, a British computer programmer, began a campaign to raise funds for the construction of an analytical engine based on Babbage's original design.

1854　George Boole publishes *The Laws of Thought*, establishing the concept of Boolean logic, which underpins digital computer logic. This concept provided the basic assumptions for creating an artificial intelligence (AI) system by breaking down the concept of what thought is and how it can be represented in a nonabstract form.

1897　Nikola Tesla demonstrates a radio-controlled boat for the U.S. military, expecting a contract for radio-controlled torpedoes. The contract was not extended in the United States, but many other nations sought to purchase such munitions for coastal and harbor defenses. Tesla's experience was typical for the period, as American military leaders had little interest in supporting expensive weapons innovations, preferring to rely on geographic isolation to provide security.

1903   Orville and Wilbur Wright launch the *Wright Flyer*, which engaged in the first sustained and controlled flight by an aircraft heavier than the surrounding air.

1918   Charles Kettering designs and tests the Kettering Bug, an unmanned aircraft that functioned as an "aerial torpedo." While not truly a drone, as it was not designed for repeated use, the Kettering Bug represented the first substantial attempt to guide an aircraft without a pilot. The control system was designed by Elmer A. Sperry, founder of the Sperry Corporation.

1929   Android robots are shown in England and the United States for the first time.

1938   Willard Pollard patents a robotic arm, dubbed the Position Controlling Apparatus. This machine opened new avenues of industrial applications, thus increasing the commercial viability of robots.

1939   The Soviet Union deploys teletanks against Finland in the Winter War of 1939 to 1940, with little operational or tactical success.

        DeVilbiss Company patents Harold Roselund's automatic spray-painting arm. This design, completed only a year after Pollard's system, demonstrated the industrial applications of purpose-built machines.

        Elektro, a humanoid robot built by Westinghouse Electric Corporation, is displayed at the New York World's Fair.

1940   John Atanasoff and Clifford Berry design the first all-electronic computer. This involves a leap forward by moving away from expensive and slower mechanical systems.

British researchers of the Ultra research group build Robinson, an electric computer designed to break German military ciphers. By that time, computers could perform tedious math functions much faster than humans and could progress through a series of combinations or permutations without repetition or error, a fundamental consideration when engaging in cryptography.

1941   Konrad Zuse builds the first programmable digital computer in Germany. This computer demonstrates that such systems did not have to be purpose built, as the addition of programming capability massively increased the flexibility of the machine.

1942   Isaac Asimov first publishes the Three Laws of Robotics in his short story "Runaround." While entirely a work of fiction, the concepts have influenced ethical discussions of the use and programming of robots.

1943   Ultra research group builds Colossus, an electronic computer that uses vacuum tubes to perform calculations. It is used to break improved German military codes. This wartime, purpose-built computer speeds up the process of deciphering messages far beyond the capability of humans to break without machine assistance, and models remain in a variety of functions until 1970.

1944   The Automatic Sequence Controlled Calculator (ASCC) is designed by Howard Aiken and built by IBM. It uses paper tape programming and vacuum tubes for calculations. The ASCC demonstrates that programming, hardware, and software are being improved simultaneously and at a far faster pace than in the recent past.

1945   John von Neumann proposes the concept of permanent stored programming for computers. Although theoretical rather than practical, the idea is hindered

1945
(*cont.*)

only by memory capacity, which will soon exponentially expand.

1946

John Eckert and John Mauchly build the Electronic Numeral Integrator and Computer (ENIAC) for the U.S. Army. It was designed to calculate artillery-firing tables but was first used to assist in the development of the hydrogen bomb. The ENIAC represents a purpose-built machine that is put to new uses unexpected by the designers, a common theme in technological development.

1947

William Bradford Shockley, Walter Brattain, and John Bardeen, engineers at Bell Laboratories, create the first transistor, making vacuum tubes obsolete and heralding the microelectronics revolution. The transistor is a major hardware advance that enables miniaturization and much cheaper, more powerful systems.

1949

W. Grey Walter demonstrates electronic autonomous robots based on biological behaviors. Copying biology, both in structure and behavior, becomes increasingly important, drawing on millions of years of biological evolution for a demonstration of efficient designs.

Maurice Wilkes at the University of Cambridge Mathematical Laboratory in England builds the Electronic Delay Storage Automatic Calculator (EDSAC), the first practical stored-program computer. This design is completed only four years after von Neumann proposed the concept.

The Binary Automatic Computer (BINAC) is designed for Northrop Aircraft Company by the Eckert-Mauchly Computer Corporation. It is the world's first commercial digital computer and shows that the government no longer has a monopoly on computer development and usage.

1950   Alan Turing proposes the Turing test of artificial intelligence, which essentially states that a machine should be considered intelligent if a human observer cannot differentiate the machine's responses to a series of questions from those of a human comparative subject.

        The Eckert-Mauchly Computer Corporation is purchased by the Remington Rand Company, and the new subsidiary produces the Universal Automatic Computer (UNIVAC). UNIVAC I is the first commercial computer, with the earliest functional model delivered to the U.S. Census Bureau in 1951.

1951   Raymond Goertz designs the first teleoperated mechanical arm, utilizing force feedback technology. It is built for the Atomic Energy Commission. Force feedback makes the use of the arm truly two way, as the operator can tell when contact is being made by the arm.

1953   The world's first remotely piloted underwater vehicle is used for undersea exploration. This demonstrates the utility of remotely-piloted vehicles (RPVs) in environments too dangerous or expensive to reach by manned vehicles, such as under the oceans, in toxic locations, or in outer space, an exploratory approach that remains important in the 21st century.

1954   George Devol builds a reprogrammable robotic arm, allowing greater flexibility in industrial applications. Once again, this innovation demonstrates a tendency to move away from single-purpose designs, making industrial adoption far more likely.

1955   General Motors installs Planetbot, a commercial robot arm utilizing polar coordinates, in its Harrison, Pennsylvania, radiator plant. The producer, Planet Corporation, had been in existence for less than one year.

1957   John W. Backus develops FORTRAN, the first high-performance programming language, for International Business Machines (IBM). It was later superseded by John McCarthy's LISP.

             The Soviet Union launches *Sputnik*, the first artificial satellite to reach earth's orbit.

1958   Jack St. Clair Kilby develops the first integrated circuit while working for Texas Instruments. This innovation opens the possibility of advances in miniaturization, and the concept of the personal computer becomes a distant possibility.

             Seymour Cray, working for the Control Data Corporation, builds the first transistorized supercomputer. A supercomputer is a machine that is at the cutting edge of processing, usually measured by the speed of calculations that it can perform.

             The Advanced Research Projects Agency (DARPA) is formed under the auspices of the Department of Defense, in reaction to the Soviet launch of *Sputnik*. It soon becomes a leading source of funding for computer and robotics research, particularly those applications of military interest.

1959   John McCarthy and Marvin Minsky found the Massachusetts Institute of Technology (MIT) Artificial Intelligence Laboratory. Artificial intelligence is a key enabling technology of truly autonomous machines, although even in 2012 no device has yet been built that can truly be considered intelligent.

1960   Theodore Harold Maiman builds the world's first laser. Soon, lasers are applied to military and communication applications.

1961   Joseph Engelberger founds Unimation Company in Danbury, Connecticut.

The first Unimate robot is installed at a General Motors assembly line in Ternstedt, New Jersey, welding die castings onto automobile chassis.

1963   John McCarthy founds the Stanford University Artificial Intelligence Laboratory. It is modeled on the MIT system, but the new location to train and inspire scholars means a greater number of innovations and a different approach to problem solving, making major technological advances exponentially more likely to occur.

1965   Gordon Moore publishes "Cramming More Components onto Integrated Circuits" in *Electronics*. In this article, he correctly predicted that the number of components per integrated circuit would double each year (Moore's Law), leading to exponential development of electronic devices and computing power. This prediction is made only seven years after the development of the integrated circuit, yet has remained remarkably prescient for over four decades.

1966   Luna 9 and Surveyor, immobile robots launched by the Soviet Union and the United States, respectively, make soft landings on the lunar surface and transmit back imagery of their surroundings to earth. Luna 9 arrives first but transmits photographs for only three days. Surveyor arrives four months later but sends engineering data, photos, and video for more than seven months.

1968   Marvin Minsky designs and builds a tentacle robot arm at the MIT Artificial Intelligence Laboratory. This is a much more complex biomimetic system than a simple arm, with inherent flexibility and utility but also with substantially more engineering challenges.

Intel Corporation is founded in Santa Clara, California. It soon comes to dominate the microprocessor market.

1969   Victor Sheinman invents the Stanford Arm, a six-axis articulated robot arm capable of following arbitrary paths in space. The development is quickly incorporated into industrial applications such as welding and assembly.

1970   Shakey is built at the Stanford Research Institute. It is the first mobile robot that can reason about its surroundings by combining multiple sensory inputs to aid in navigation. It moves incredibly slowly, limited by processing power and the need to map its environment. However, it is capable of acting independently of any human control and in a nonprogrammed situation.

The Soviet Union lands Lunokhod 1, the first remote-controlled rover to reach the lunar surface. This alternate approach to lunar exploration demonstrates that such missions need not be manned. It is less expensive and less dangerous than the American *Apollo* missions, and the initial device remains functional for more than 10 months.

1971   Intel Corporation develops the first commercial microprocessor, dubbed the 4004, a 4-bit central processing unit contained on a single chip. It is the first use of silicon gate technology, allowing the then-revolutionary speed of 740 kHz.

1972   The Teledyne Ryan Lightning Bug unmanned aerial reconnaissance vehicle uses a television camera to broadcast live images to a controller aircraft.

1973   The Department of Artificial Intelligence at the University of Edinburgh completes Freddy II, a robot capable of assembling complex objects from piles of components. While this robot does not possess true artificial intelligence, it is a major leap forward in the construction of robots demonstrating complicated behavior and abstract reasoning.

Wabot-1, a humanoid robot, is constructed at Waseda University in Tokyo. It communicates verbally in Japanese and uses artificial eyes and ears to measure distances and directions to objects. This device shows a major shift toward android systems, most prevalent today in Japan.

1974   The Robotics Industries Association (RIA) is founded in Ann Arbor, Michigan.

David Silver designs and builds the Silver Arm, an artificial arm utilizing pressure sensors to provide feedback. Pressure sensors are far more precise than a force feedback system, allowing much more minute control over the arm.

1975   Automatix Company is founded, specializing in industrial robot production of systems with machine vision.

1976   Viking 1 and 2 land on the surface of Mars, and use robot arms to take soil samples and perform experiments. These are the first robots to visit the surface of another planet. They continued to function for more than six and three years, respectively, far longer than mission expectations.

Stephen Wozniak and Steven Jobs found Apple Computer Corporation. The company soon specializes in home computers with user-friendly systems, and today remains a major competitor to personal computer (PC) designs.

1977   George Lucas releases *Star Wars*, a film that includes two robots, C-3PO and R2-D2, as key heroes. In the film, the term "droid" is coined to refer to an autonomous robot possessing artificial intelligence.

Voyager 1 and 2 are built and launched. These robotic space probes are sent to examine the four gas giant planets of the solar system. Both are still in operation,

1977 (*cont.*)	far beyond the range of Pluto's orbit, and are the two farthest human-created objects from Earth. Despite being in operation for more than 34 years, these spacecraft continue to receive and respond to commands from Earth.
1978	Unimation develops the Programmable Universal Machine for Assembly (PUMA).
1979	Hans Moravec builds the Stanford Cart, a robot that relies on stereovision to navigate through a room full of obstacles. Faster than Shakey, this robot compares images of the environment to stored information, mimicking biological vision in a crude fashion. Raj Reddy founds the Robotics Institute at Carnegie Mellon University.
1981	IBM releases its first personal computer (PC). As computers begin to enter homes, amateur users begin to experiment with their machines, allowing innovation to occur outside of formal research channels.
1982	The Israel Defense Force (IDF) deploys RPVs against Syrian forces in Lebanon, providing effective reconnaissance and target designation for Israeli ground forces.
1984	Researchers at Waseda University in Tokyo design Wabot-2, a humanoid robot capable of reading musical scores with artificial eyes and playing songs on an electronic organ. These machines are reminiscent of de Vaucanson's musicians of two centuries earlier but much more complex. They can interpret new data rather than being limited by preprogrammed scores. Stanford Artificial Intelligence creates Flakey, a mobile robot that uses stereovision and is capable of recognizing and following individual humans. The recognition system predates later weaponized target recognition approaches.

DARPA launches the Autonomous Land Vehicles (ALV) initiative, designed to create robotic transport vehicles for the U.S. military.

Apple Corporation releases its first Macintosh desktop computer, a major competitor to IBM's PC systems.

The Terminator, a film directed by James Cameron, debuts in the United States. It features Arnold Schwarzenegger as a futuristic robot determined to kill the protagonists of the film.

1985   Kawasaki Heavy Industries begins to produce robots. This major engine producer's interest in robots shows the amount of capital involved in robotic developments.

1986   Honda Motor Company begins research and development of humanoid robots. Like Kawasaki, Honda is a major engine manufacturer that has recognized robotics as a vital industry in Japan.

Carnegie Mellon's Robotics Institute designs its first NavLab (Navigation Laboratory) robot capable of autonomously driving a truck.

The RQ-2 Pioneer is adopted for service in the U.S. and Israeli military forces. It is built by AAI Corporation, formerly Aircraft Armaments, Inc.

1988   Danny Hillis reveals his Connection Machine, a parallel processor capable of more than 65,000 simultaneous computations. This device is an order of magnitude faster than previous designs, demonstrating a revolutionary leap forward in computer architecture.

1989   MIT Robotics Institute demonstrates Genghis, a fast-moving insect-like robot using bottom-up processing and available for mass production. This robot is

1989 (*cont.*)	small, cheap, and reflects Rodney Brooks's vision of robots engaged in swarm behavior.
	Researchers at the Jet Propulsion Laboratory (JPL) complete a wheeled planetary rover, Tooth. This will eventually fold into a series of Martian rover systems.
1990	Howard Paul and William Bargar demonstrate Robodoc, a robotic arm used to perform a hip replacement surgery on a dog. The Robodoc allows a surgeon's movements to be duplicated on a much smaller scale, allowing surgical procedures to be much less invasive.
1992	Robodoc performs a human hip replacement. It has been in operation on animals for less than two years but is such an important advance that it is accelerated into human trials.
	Japanese academic, government, and industrial institutions team up to create the Humanoid Project.
1993	Rodney Brooks commences construction of Cog, a robot designed to learn new information in the same manner as a human child. Brooks has created the world's first biomimetic AI system, which starts with the fundamentals of acquiring knowledge but then acquires new information at an exponential pace.
1994	Carnegie Mellon's Robotics Institute creates Dante II, a robot that enters Mt. Spurr, an active volcano in Alaska, and samples the gases found within the crater.
	General Atomics debuts the RQ-1 Predator, which makes its maiden flight as an intelligence, surveillance, and reconnaissance (ISR) platform.
1995	Waseda University in Tokyo demonstrates Hadaly-1, a humanoid robot designed to be integrated into human society.

The RQ-1 Predator is accepted into military service and makes its first combat deployment to assist in monitoring conflicts in the former republic of Yugoslavia.

1996   David Barrett builds RoboTuna, a biomimetic robot that emulates a swimming fish. This propulsion method has potential military applications, as it is both efficient and nearly silent.

Honda Motor Company demonstrates Prototype 2 (P-2), a humanoid robot capable of walking, climbing stairs, and carrying objects.

1997   Sojourner conducts semiautonomous exploration of the surface of Mars, choosing its own path to navigate around obstacles. Although intended for a week of operations, it actually continues to function for three months.

Deep Blue, a supercomputer built by IBM, defeats World Chess Champion Garry Kasparov in a six-game match. Kasparov is considered one of the greatest chess minds in history, and a computer capable of playing chess at such a high level has been a goal of programmers for decades.

NEC Corporation begins development of the PaPeRo remote presence personal robot. Such companion robots potentially anthropomorphize machines, making them more accepted in human homes. A remote presence robot is designed as a sophisticated communication device, allowing a distant operator to not only see and hear others, but to physically interact with them.

The first RoboCup robotic soccer match is held in Japan. It is the first of a host of robotic "sports."

The U.S. National Aeronautics and Space Administration (NASA) begins to design and construct

1997 (*cont.*)	Robonaut, a humanoid meant to assist astronauts engaged in spacewalks. By the mid-21st century, it might replace humans entirely in the performance of these dangerous missions.
1998	Cynthia Breazeal begins construction of Kismet at MIT. It is a robot designed to use vision and speech to recognize human social cues, including emotions, and it has enough facial characteristics to simulate its own emotions, allowing studies on the interaction between humans and machines.
	The Furby interactive robot is introduced to the international toy market. This device simulates emotions, can communicate with other Furbies through infrared ports, and gradually "learns" English from its owners. In reality, these devices were simply programmed to utilize more English words, simulating a language development that did not exist. For this reason, the misperception that they repeated words said in their presence caused a minor panic in intelligence agencies, which banned the popular toys out of fear that they might create a leak of classified information.
1999	Sony introduces the AIBO, a robotic dog capable of following spoken commands.
2000	The U.S. Congress calls for one third of all aircraft in the American operational deep strike force aircraft fleet to be unmanned by 2010, and one third of all operational ground combat vehicles to be unmanned by 2015.
	Honda reveals ASIMO, a humanoid robot that can walk, run, recognize faces, and communicate with humans.
	A Triton XL marine remotely operated vehicle assists in efforts to stop leakage from the sinking oil tanker *MV Erika*.

iRobot Corporation reveals My Real Baby, a robotic doll that uses animatronics to simulate emotional responses. The company, founded by Rodney Brooks, Colin Angle, and Helen Greiner, commercializes advances developed in the MIT laboratories.

The RQ-8 Fire Scout, designed by Northrop Grumman, makes its first flights.

2001 The RQ-4 Global Hawk makes the first autonomous nonstop flight over the Pacific Ocean in 22 hours. This flight demonstrates the range and endurance of the new reconnaissance platform. The autonomous operation reduces the chances of operator error and simultaneously lowers the aircraft's bandwidth requirements.

The Space Station Remote Manipulator System (SSRMS), also known as Canadarm2, is deployed to assist in the construction of the International Space Station (ISS).

Fujitsu Animatronics demonstrates the Humanoid for Open Architecture Program (HOAP-1), a commercial humanoid robot with USB interface to allow easy modifications.

Dr. Jacques Marescaux performs the first telesurgery, using a Zeus surgical robot to remove a French patient's gallbladder from New York. Telesurgery allows experts to serve patients on a global scale, increasing the accessibility of sophisticated medical techniques around the world.

Robots are used in search and rescue efforts at the World Trade Center site in the aftermath of the September 11, 2001, attacks. They do not suffer from fatigue or boredom and can operate in potentially harmful environments that would hinder human rescue efforts.

2001
(*cont.*)
Predator UAVs are armed with AGM-114 Hellfire missiles to test the feasibility of using them as hunter-killer aircraft. When the tests prove successful, the newly armed variants are renamed MQ-1 Predators.

General Atomics Aviation Services, Inc., tests the first prototypes of the Predator B, later renamed the RQ-9 Reaper.

QinetiQ is founded under the control of the United Kingdom's Ministry of Defence. A defense industry consortium, it soon became one of the largest manufacturers of military robots and UAVs in the world, and acquired a series of other defense corporations.

2002
iRobot Corporation releases the Roomba robotic vacuum cleaner. In less than a decade, more than six million units were sold worldwide.

A Predator UAV armed with Stinger missiles engages an Iraqi MiG aircraft in the first UAV versus manned fighter engagement in history. The Predator is shot down, while the MiG sustains no damage.

2004
The Centibots Project demonstrates the ability of 100 small autonomous robots to cooperate in mapping, searching, and guarding an unknown environment. This demonstrates the feasibility of Rodney Brooks's vision of small, cheap robots operating in swarms.

The first DARPA Grand Challenge competition is held in the Mohave Desert. None of the entrants finished the course, with the most successful entrant, Carnegie Mellon's Red Team, completing only 11 kilometers of the 240-kilometer course.

The robot rovers Spirit and Opportunity land on Mars and begin mapping the surface. Spirit became been stuck in soft soil in 2009 and stopped transmitting in 2010. Opportunity continues to function and collect data on Mars and transmit it back to Earth.

British defense consortium QinetiQ acquires Foster-Miller and continues to produce the highly successful line of robots designed by the American company.

2005   Researchers at Cornell University demonstrate the first robot capable of assembling a copy of itself without human intervention, although it is incapable of manufacturing the components necessary for assembly of the copy.

Wakamaru, a Japanese domestic robot produced by Mitsubishi Heavy Industries, goes on sale. It is designed as a companion and assistant for people with disabilities and older adults, and has Internet connectivity as well as limited speech recognition capabilities. It utilizes a Linux operating system and is available for an initial sale price of $14,000.

Five teams complete DARPA's Grand Challenge off-road course. The competition is won by Stanford University's Stanley, which completes the 212-kilometer course in under seven hours. Three other teams finish less than an hour behind Stanley, and a fifth completes the course in 13 hours.

2006   Stanford researchers demonstrate the Starfish, a robot that can self-model and relearn to walk after sustaining damage. The robot demonstrates a heretofore unseen level of adaptability to both battle damage and unpredictable terrain, leading to a host of potential military applications.

Unmanned platforms assist in tracking the location of Abu Musab al-Zarqawi, leader of Al Qaeda in Iraq. He is killed in a piloted airstrike.

2007   Google Corporation announces the Lunar X Prize competition, an award of $30 million for the first private company that lands a rover on the lunar surface that is capable of sending images back to Earth.

DARPA's Urban Challenge is held at George Air Force Base in Victorville, California. Carnegie Mellon's Boss wins the competition, completing the 100-kilometer urban course in just over four hours. Five other teams finished the race.

RQ-9 and MQ-9 Reapers are deployed to Afghanistan for the first time, greatly enhancing the ability to undertake hunter-killer aircraft missions with UAVs.

RQ-1 and MQ-1 Predators reach 250,000 hours of total flight time.

Foster-Miller's TALON SWORDS variant is deployed to Iraq for the first time.

2008   Confirmed drone strikes in Pakistan rise from 3 in 2007 to 16 separate incidents, killing at least 185 individuals.

Annual defense spending on robotic systems tops $3 billion.

Foster-Miller announces it has delivered 2,000 TALON robots for service in Iraq and Afghanistan.

2009   RQ-1 and MQ-1 Predators reach 500,000 hours of total flight time.

The General Atomics Avenger, a turbojet-powered improvement upon the MQ-9 Reaper, makes its first flights.

2010   RQ-1 and MQ-1 Predators reach more than a million hours of total flight time.

Honda announces the successful testing of a Brain Machine Interface prototype that will theoretically allow a robot to be controlled solely by thoughts.

2011   iRobot PackBots and Warriors are sent into the flooded Japanese Fukushima Daiichi nuclear plant to assist in cleanup operations.

The last planned MQ-1 model is delivered to the U.S. Air Force.

Unmanned aircraft are used for both reconnaissance and air strikes in Libya and Yemen.

Drone aircraft assist in finding and coordinating the raid on the compound where Osama bin Laden is hiding in Abbottabad, Pakistan.

Drone aircraft strike kills Anwar al-Awlaki, American citizen and senior leader of Al Qaeda in the Arabian peninsula. The attack prompted a series of lawsuits arguing that the CIA had violated his civil rights through the targeted killing. Two weeks later, a drone strike kills his son.

The RQ-170 Sentinel, one of the most sophisticated and capable UAVs in the American inventory, is lost in airspace over Iran. The Iranian government later reported it had captured the aircraft intact.

Boeing ScanEagles cooperate with a Procerus Unicorn to demonstrate swarm behavior using dissimilar UAVs.

American UAVs are dispatched to assist NATO operations in Libya.

2012   Despite previous denials, the U.S. government confirms an active drone strike program in Pakistan and refuses to stop attacks on Al Qaeda militants despite protests by the Pakistani parliament.

2012
(*cont.*)

Iran announces it has successfully extracted the data contained within the captured RQ-170 Sentinel and decrypted its information. The Iranian government also announces it is in the process of reverse-engineering the Sentinel to produce copies.

Unmanned aircraft are dispatched to Uganda to assist in the search for Joseph Kony, leader of the Lord's Resistance Army terror organization.

The robot rover Curiosity lands on Mars and commences investigations into the conditions on the Martian surface, including an examination of whether the Martian environment ever supported life.

5

Biographical Sketches

To truly understand the field of military robotics, it is necessary to examine the contributions of individuals from many walks of life. In particular, those who contribute to the advancement of the field of robotics, even if they show no interest in creating military models, have a major effect on the creation of war machines. Likewise, individuals who make the fundamental decisions regarding the procurement and utilization of robots in warfare set the course of military robotics for decades to come.

Asimov, Isaac (1920–1992)

Isaac Asimov was born Isaak Yudovich Ozimov in 1920 in Petrovichi, Byelorussia. His family immigrated to Brooklyn, New York, in 1923. He was best known as a science fiction author, although he earned a doctorate in biochemistry from Columbia University in 1948 and published a substantial number of nonfiction works while serving on the faculty of the Boston University School of Medicine.

Asimov is credited with coining the terms "robotics" and "positronic," both in fictional works. In the field of robotics, he is best known for imagining his Three Laws of Robotics, a set of rules incorporated into virtually all of his imagined robots. The Three Laws establish an overarching operating system, governing the behavior of autonomous machines, and are often referred to by modern robotic developers. The First Law states that a robot may not injure a human being or, through inaction, allow a human being to come to harm. The Second Law commands that a robot must obey any orders given to it by human beings, except where such orders would conflict with the First Law. The Third Law orders that a robot must protect its own existence as long as such protection does not conflict with the First or Second Law. While some designers see the concept of these laws as a means to ensure the safe development of robots that will remain completely under the control of their creators, others point out that such laws would preclude the development of military robots.

Sources

Asimov, Isaac. 1994. *I. Asimov: A Memoir.* New York: Doubleday.

Asimov, Isaac and Stanley Asimov. 1995. *Yours, Isaac Asimov: A Lifetime of Letters.* New York: Doubleday.

Gunn, James E. 1982. *Isaac Asimov: The Foundations of Science Fiction.* New York: Oxford University Press.

Brooks, Rodney (1954–)

Rodney Brooks was born in Adelaide, South Australia, in 1954. He attended Flinders University in Adelaide, where he completed his bachelor's (1975) and master's degrees (1978) in pure mathematics. He studied at Stanford under John McCarthy, examining an artificial intelligence problem for his doctoral dissertation. He graduated in 1981. In 1984, he accepted a position at the Massachusetts Institute of Technology (MIT) as a professor of robotics, working in the Mobile Robotics Laboratory.

Brooks transformed the study and development of artificial intelligence by arguing that the top-down processing systems that dominated the field in the 1970s and 1980s created an inherently flawed approach to artificial intelligence. He believed that attempting to supply a machine with a complex representation of the world served only to slow the machine's ability to interact with its environment. Rather, he argued that robots should be programmed in a bottom-up fashion, with the emphasis placed on interactions and behavior, not on an internal understanding of the robot's surroundings. This vastly simplified the problem of programming, while allowing for surprisingly complex robotic behaviors. Essentially, Brooks allowed his creations to learn from their environment, rather than trying to give them every piece of information possible prior to activation. Much of his design philosophy has been drawn from biological examples, particularly from the insect world. Essentially, Brooks programs his creations with "instincts" rather than world-modeling, resulting in very simple compulsions that can create extremely complex behaviors. For example, an instinct to wander and map terrain, coupled with obstacle avoidance and a desire to maintain distance from other identical robots, can map an enormous territory in a very short period.

In 1997, Brooks became the director of the MIT Artificial Intelligence Research Laboratory, supervising more than two dozen doctoral students in robotics and artificial intelligence research. From 2003 until 2007, he served as the director of the MIT Computer Science and Artificial Intelligence Laboratory. He is now the Panasonic Professor of Robotics (emeritus) at MIT. Brooks also cofounded the iRobot Corporation, in partnership with Colin Angle and Helen Greiner, in 1990. The company specializes in autonomous robots for home usage, although it has increasingly won contracts to design and produce robots with military applications. The best-known iRobot products include the Roomba floor vacuum, first released in 2002, and the PackBot multipurpose utility robot, first deployed with U.S. troops in Afghanistan in 2002. The company continually designs and produces both commercial and military robots, and remains a leader in technological development of robot systems.

Sources

Brooks, Rodney Allen. 2002. *Flesh and Machines: How Robots Will Change Us*. New York: Pantheon Books.

Brooks, Rodney. 2012. *Rodney Brooks: Roboticist*. Retrieved from http://people.csail.mit.edu/brooks/.

Bush, George W. (1946–)

George W. Bush was the 43rd president of the United States. He was born in New Haven, Connecticut, in 1946. After completing a bachelor's degree in history at Yale University in 1968, he entered the Air National Guard on active duty as a fighter pilot. Upon completion of his service, he entered the Harvard Business School, earning an M.B.A. in 1975. He then entered a business career, largely based in Texas, which centered around the oil industry. In 1994, he defeated the incumbent Ann Richards in the gubernatorial race in Texas. In 2000, Bush ran for the presidency, defeating Vice President Al Gore in a hotly contested election marred by controversy.

Bush was in office during the deadliest terror attacks in American history, the September 11, 2001, airliner hijackings. These attacks forced him to confront a terrorism threat that had previously been considered a problem faced by other nations and propelled him to commit American military forces to an invasion of Afghanistan with the goal of destroying Al Qaeda. While American forces were able to quickly overthrow Afghanistan's ruling party, the Taliban, and destroy the Al Qaeda bases located in the hinterlands of the country, finding Al Qaeda's leader, Osama bin Laden, proved far more difficult. America's conventional forces were not well designed for this type of a campaign and came to increasingly rely on overwhelming firepower against a growing insurgency.

In 2002, the president began to make the case that Iraq represented another substantial threat to the United States, in part due to its leader's support of terror organizations, and in part due to its pursuit of weapons of mass destruction (WMDs). In March 2003, American forces led a coalition of nations determined to eliminate the Saddam Hussein regime in Iraq and destroy any remaining WMDs within the country. Although the coalition quickly overwhelmed the Iraqi defenders and removed the Hussein administration, they were not welcomed as liberators by many Iraqis, and an insurgency against Western occupation soon began to arise in Iraq.

Although unmanned aerial vehicles had been used in foreign locations by the U.S. military prior to the Bush administration, their usage became much more common in Afghanistan and Iraq. Further, the first armed variants of the most common unmanned aerial vehicle (UAV) platforms began to operate in the skies over each location. In 2003, RQ-1 Predators began to launch airstrikes with Hellfire air-to-ground missiles, creating an intelligence, surveillance, and reconnaissance (ISR) platform capable of an almost instantaneous response against fleeting targets. The program proved so successful that the administration pushed to increase the procurement of armed UAVs by both the military and the intelligence services.

As the insurgency expanded in Afghanistan and Iraq, improvised explosive devices (IEDs) became the deadliest weapon in the insurgent arsenal. To assist in disabling and disarming these weapons, the Pentagon sought to deploy increasing numbers of ground robots that could be remotely operated, keeping explosive ordnance disposal (EOD) teams out of lethal range. Like the armed UAVs, the first experiments proved so successful that procurement budgets were rapidly expanded, as were development programs to create an increasing autonomous capability within the armed forces.

President Bush authorized the first UAV strikes not only in Afghanistan and Iraq, but also in Pakistan and Yemen, greatly expanding the definition of what could be considered a combat zone. His avowed Bush Doctrine, that the United States would reserve the right to act preemptively against any perceived threat to the security of the United States, also paved the way for many future attacks on sovereign nations. The most common form of such strikes has long been through aerial operations, but the possibility of losing American pilots has served as a potential deterrent against such attacks. With the rise of remotely piloted strike aircraft, the ability to attack with impunity has increased, a fact that has contributed to President Barack Obama's decision to act in multiple locations against terror targets, primarily using armed UAVs.

Sources

Bush, George W. 2010. *Decision Points*. New York: Crown Publishers.

Cheney, Richard B. 2011. *In My Time: A Personal and Political Memoir*. New York: Threshold Editions.

Rice, Condoleezza. 2011. *No Higher Honor: A Memoir of My Years in Washington*. New York: Crown Publishers.

Cebrowski, Arthur (1942–2005)

Arthur Cebrowski was a career officer in the U.S. Navy, best known for leading the Office of Force Transformation in the U.S. Department of Defense. Cebrowski pushed for the

American military to radically reorient itself in the post–Cold War period, believing that the force was ill prepared to face the emerging threats that would confront the military in the 21st century. Cebrowski coined the term "network-centric warfare" to describe his vision for the transformation of the armed forces.

Cebrowski was born in Passaic, New Jersey, in 1942. He attended Villanova University as a Naval Reserve Officer Training Corps student, graduating in 1964 with a degree in mathematics. Upon graduation, he accepted a naval commission and commenced aviation training. Upon earning his pilot's wings, Cebrowski received orders to deploy to Vietnam, where over the course of two tours totaling three years, he flew 154 combat missions and won a substantial number of decorations, including the Distinguished Service Cross. In 1981, he was selected as an inaugural member of the Naval War College's Strategic Studies Group, a small class of mid-grade officers charged with studying the long-term strategic posture of the U.S. military. Most of the members of the class, including Cebrowski, eventually rose to flag rank.

Most of Cebrowski's career was spent in command billets within the fleet, rather than staff positions at the Pentagon or other shore locations. He commanded the *USS Midway* during the Persian Gulf War and a carrier battle group as a rear admiral after the war. In 1998, Vice Admiral Cebrowski received an appointment to serve as the president of the Naval War College, a position he held until his retirement in 2001. As the leader of the Navy's graduate school, Cebrowski focused on the major technological, doctrinal, and strategic problems confronting the Navy. He argued for the development of smaller, more flexible surface warships, including the littoral combat ship (LCS), the Streetfighter stealth warship, and the "corsair" small carrier. He believed that a large number of small vessels presented more flexible options to the political and military leaders of the nation, who he expected to confront a host of smaller conflicts in the 21st century. Key to the success of these vessels was the development of robust information-sharing networks. To Cebrowski, the new model of conflict revolved around the collection, analysis, and utilization of information in the fastest and most efficient manner possible. He believed that decentralized networks of independent actors, working in conjunction with a centralized information database, could respond most flexibly to any future threat. At the same time, smaller and more nimble vessels would be less vulnerable to enemy action and would provide too many targets for an enemy to launch a devastating surprise attack.

In 2001, Cebrowski retired from active duty but continued to serve his country by confronting the strategic problems facing the Pentagon. Secretary of Defense Donald Rumsfeld asked Cebrowski to develop and lead the newly created Office of Force Transformation. The organization combined the analytical ability of a think tank with direct access to the nation's political leadership. Rumsfeld wished to cut through some of the bureaucratic inertia of military change as the American military shifted from the large, heavy formations of the past to the faster, lighter units of the current era. In addition, the acquisition timeline for new military technology has become so long that newly created machines are obsolete before they reach full production, and he hoped that the new organization would be able to identify key platforms for development in a much faster manner.

Over his four years in the position, Cebrowski sought to change attitudes of both civilians and military personnel regarding the application of military force. He believed that changing the culture of the organization began at the ground level but also thought the military's professional military education (PME) system offered a unique way to transform officers' thinking. He took a particular interest in the role of emerging technologies, pushing military officials to devote more attention and resources to unmanned systems and cyber capabilities, rather than simply following the force development methods of the past, and as such, is directly responsible for the rise in the number of robots and drones within the military inventory.

Sources

Blaker, James R. 2007. *Transforming Military Force: The Legacy of Arthur Cebrowski and Network Centric Warfare.* Westport, CT: Praeger Security International.

Singer, Peter W. 2009. *Wired for War.* New York: Penguin.

Gates, Robert M. (1943–)

Robert M. Gates served as the secretary of defense under the administrations of President George W. Bush and President Barack Obama. Previously, he served in the administration of President George H. W. Bush as the director of the Central Intelligence Agency (CIA). In each capacity, he was instrumental in the development and deployment of autonomous military weaponry.

Gates was born in Wichita, Kansas, in 1943. He completed a bachelor's degree in history at the College of William and Mary in 1965, followed by a master's degree in history from Indiana University in 1966 and a Ph.D. in 1974 from Georgetown. His doctoral dissertation examined the relationship between the Soviet Union and China, and identified a number of fractures between the communist countries. In 1966, Gates joined the CIA, which sponsored him for officer training in the U.S. Air Force. After two years as a military intelligence officer, he returned as an analyst to the CIA. In 1974, he joined the staff of the National Security Council but returned to the CIA in 1979 and gradually worked his way through the organization. In 1991, he was nominated and confirmed as the director of the CIA, the only career officer to rise from an entry-level position to the head of the organization.

After retiring from the CIA in 1993, Gates moved into academia, serving first as the dean of the George Bush School of Public Policy and Administration at Texas A&M University before becoming the president of the school in 2002. In 2006, he was confirmed as the secretary of defense, succeeding Donald Rumsfeld in the position. He proved a powerful and active secretary, removing several key political leaders for their roles in scandals, including the secretaries of the Army and Air Force. He also demonstrated a willingness to cancel expensive legacy programs that did not provide the hardware needed by the 21st century military, focusing instead on unmanned systems and greater automation within the armed forces. In 2008, President-elect Obama announced that Gates would continue to serve in the same post after the president's inauguration, making Gates the first secretary of defense to serve two presidents of different political parties.

Sources

Gates, Robert M. 2007. *From the Shadows: The Ultimate Insider's Story of Five Presidents and How They Won the Cold War.* New York: Simon & Schuster.

Spartacus Educational. *Robert Gates.* Retrieved from http://www.spartacus.schoolnet.co.uk/MDgatesR.htm.

Greiner, Helen (1967–)

Helen Greiner is a pioneer within the robotic industry, both as a designer of robots and the cofounder of iRobot Corporation and CyPhyWorks. Greiner studied mechanical engineering and computer science at the Massachusetts Institute of Technology, working under the tutelage of Rodney Brooks. She worked for the National Aeronautics and Space Administration (NASA) Jet Propulsion Laboratory, the MIT Artificial Intelligence Laboratory, and California Cybernetics before founding iRobot in 1991.

Greiner was born in London but raised in Southampton, New York. Showing an early interest in science, by age 10 she decided on a career in robotic design, inspired in part by seeing the *Star Wars* character R2-D2. By her teenage years, Greiner had learned to write programs for the family's home computer, using it to give orders to remote-controlled toys.

After completing her master's degree at MIT, Greiner was approached by Brooks and fellow MIT alumnus Colin Angle and asked if she would join in a venture to construct practical, commercial robots. Greiner had long believed that robots should move beyond the laboratory and into society, through home use and a much wider series of applications in industrial manufacturing. The three began their company with their own funds, working from Angle's apartment to build robots for academic researchers. Greiner served as president of the new company. Soon, they drew the interest of the U.S. Department of Defense, as the Office of Naval Research contracted the group to build a robot capable of detecting and disabling maritime mines. Greiner did much of the design work, basing the robotic system on the movements and behavior of the ghost crab.

As president of the company, Greiner took the lead in negotiating further defense contracts. In 1995, she won a bid to design a small scouting platform. This robot, eventually named the

PackBot, needed to be portable, maneuverable, and capable of navigating difficult terrain in both urban and wilderness settings. The initial design was well received and useful to troops in the field, ensuring the company would receive a series of follow-on orders for the system. Rather than being content with this lucrative revenue stream, Greiner pushed to improve the robot's design, based largely on feedback from troops in the field, making iRobot popular with the military forces using its equipment and establishing the company's reputation for having a much better grasp of how end users perceive its robots than the norm for similar companies.

Although defense contracting remained a fundamental part of the iRobot corporate mission, the company also sought to move into the consumer market, recognizing that the civilian customer base represented an enormous potential for future products. In 2000, iRobot and Hasbro partnered to release My Real Baby, a robotic doll with cutting-edge artificial intelligence technology that allowed it to respond to interactions with users and to gradually learn from its owners. Considered a failure in the toy industry, My Real Baby still sold more than 100,000 units, and demonstrated that the public was willing to purchase and use robotic technology, given the correct applications.

In 2002, iRobot introduced the Roomba, a robotic vacuum cleaner designed to clean floors without any human assistance. It runs on rechargeable batteries and can be programmed to clean rooms on a predetermined schedule or on demand. The Roomba determines its own pattern, navigating around furniture and other obstacles, and returning to its charging dock upon completion or when its batteries run low. Sensors in the latest models of the machine can determine if an area is particularly soiled and hence requires more attention. Roomba is the highest-selling computer robot in history, with more than 5 million units sold to date. It has been followed at iRobot by robotic floor moppers, pool cleaners, gutter cleaners, and lawn mowers.

Greiner remains an aggressive visionary, seeking to increase the utilization of robots in virtually every aspect of human society. She believes that within a decade, robots will be capable of such sophisticated interactions with humans that they will be capable of operating as both personal assistants and companions. In her vision, robots will assume responsibility for virtually every dangerous job in human society, freeing humans to devote their energies and interests to other tasks and leading to a true social revolution based on robotic technology. Her connections to the U.S. military led to the 2008 founding of CyPhyWorks, a robotics company that specializes in developing unmanned aerial vehicles for both military and commercial customers.

Source

Encyclopedia of World Biography. 2012. *Helen Greiner.* Retrieved from http://www.notablebiographies.com/news/Ge-La/Greiner-Helen.html.

Kurzweil, Raymond (1948–)

Raymond Kurzweil is an American scientist and author who specializes in computer-human interfaces and predictions about the future of technological development. He has contributed to innovations in artificial intelligence, optical character recognition, speech synthesis, and speech recognition. Many of his scientific contributions have provided key capabilities for the improvement of robotic designs, particularly of robots designed to interact directly with humans. His published works, including both fiction and nonfiction, have influenced the public's understanding about both the current level of achievements possible in robotic design and the likely advances in robotics in the near and intermediate future.

Kurzweil was born and raised in New York City. As a teenager, he learned of the new field of computer science and immediately began to demonstrate an aptitude for programming. By the end of high school, Kurzweil had designed and written a program that analyzed classical music and searched for patterns in the compositions. The program then created entirely new works of a similar style. This program won the top prize at the International Science Fair of 1965 and caught the attention of computer specialists at the Massachusetts Institute of Technology (MIT).

Kurzweil earned his bachelor's degree from MIT in 1970, majoring in computer science and literature. While working on the degree, he wrote a program that compared thousands of colleges to questionnaires completed by high school students. It offered suggestions for the best matches between students and schools. Kurzweil sold the rights to the software to publishing giant Harcourt, Brace & World, generating the funds necessary to start his own company, Kurzweil Computer Products, in 1974. Kurzweil's new company focused on the problem of optical character recognition (OCR). He wanted to create a scanning system that was not restricted to a single difficult-to-read font. Within two years, the company produced not only a multifont OCR scanner but also the Kurzweil Reading Machine, a device that scans pages electronically and reads them aloud. This invention showcased a lifelong passion, that of making communication technology more accessible to individuals with disabilities. In this case, the machine revolutionized the accessibility of books for the blind. At the same time, it opened up entirely new avenues for computers to absorb information from existing sources.

In 1978, Kurzweil Computer Products made its OCR software commercially available. Soon, archives and publishers began using it to create digital databases of material. Where old newspapers had once occupied enormous file cabinets filled with microfilm, they now began to occupy computer hard drives. In the 1980s, Kurzweil began work on an early speech recognition program. It relied on pattern recognition and once again offered a new approach to the future interactions of humans and machines. The emphasis on pattern recognition led to the 1999 debut of a hedge fund run entirely by computers that spot microtrends in financial markets.

Many of Kurzweil's published works engage in futurism, predicting and promoting the technological advances of the coming decades. His ideas have often proven controversial but also prescient on a number of occasions. He sees technological innovation following some of the same patterns as biological evolution. However, he also considers computers and their processing power to be the foundation of major advances in many other scientific disciplines. He expands on Moore's Law, arguing that computing power doubles every 18 months. As a result, revolutionary advances in other fields are occurring at an exponential pace. He believes that in the near future, a "singularity" will occur, a point in time in which all future human civilization will be so radically different from the past as to be completely unrecognizable.

Kurzweil has offered many predictions in the past two decades, and in some areas, he has been remarkably accurate. In his 1990 book *The Age of Intelligent Machines*, he correctly foresaw the massive growth of the Internet, including predictions about the worldwide adoption of the technology, the wireless means of access, and the linkages between enormous databases of purely digital media. Kurzweil has recently become particularly interested in the future potential of nanotechnology. He argues that while this field of study could have enormous positive effects on human civilization, including advances in medical technology that might expand life spans indefinitely; changes in our interaction with the environment that might prevent or reverse global warming; and power generation that could remove any reliance on fossil fuels. However, nanotechnology could also prove disastrous, as it could lead to new forms of devastating weapons; perfectly engineered plagues; or swarms of self-replicating machines that destroy the planet's resources in a mindless quest for building materials. He urges government leaders not to try to block technological progress but instead to facilitate and regulate beneficial developments.

Kurzweil believes that humans will eventually transcend their biology through the use of technology. He sees human cloning as an inevitable development of the near future, including the possibility of harvesting DNA from long-deceased individuals. Such a process could theoretically create a biological copy of a human, but it would not bring the dead back to life. Nanorobots, though, might harvest the recollections of the cloned individual held by their closest family and friends. Theoretically, these memories might be implanted in the clone's brain, creating a new individual that closely approximates the way the original person was remembered.

Unsurprisingly, Kurzweil's ideas have provoked harsh criticism in both the scientific community and the public at large. Some find the notion of the singularity to be ludicrous and argue that technology advances by gradual evolution; even if it is accelerating, there will not be one moment when everything changes. Others believe Kurzweil is mostly correct about the potential of technology but believe it will destroy civilization rather than improve it. A third set of critics object to his vision of the future on philosophical grounds, arguing that even the pursuit of such advanced technology will erode the fabric of humanity, regardless of whether the development is successful. Despite the criticism, Kurzweil remains successful as both a corporate developer of technology and a literary interpreter of its ramifications, and continues to contribute in both fields at a torrid pace.

Sources

Kurzweil, Ray, Jay Wesley Richards, George F. Gilder, et al. 2002. *Are We Spiritual Machines? Ray Kurzweil vs. the Critics of Strong A.I.* Seattle: Discovery Institute Press.

Ptolemy, Barry. 2011. *Transcendent Man: The Life and Ideas of Ray Kurzweil.* DVD Video. New York: Docurama.

McCarthy, John (1927–)

John McCarthy was born in Boston, Massachusetts, on September 4, 1927. He majored in mathematics at the California Institute of Technology, where he completed a bachelor's degree in 1948. Just three years later, he received his Ph.D. in mathematics from Princeton University, working under the tutelage of Solomon Lefschetz. McCarthy quickly became interested in the nascent field of computer science and saw mathematical logic as the rational approach to create programming for autonomous machines. He coined the term "artificial intelligence" (AI) in 1955 to describe a machine possessing the reasoning capabilities of a human being. After 11 years of teaching at a variety of top-notch universities, including Princeton, Dartmouth, and the Massachusetts Institute of Technology (MIT; where he established the Artificial Intelligence Laboratory), McCarthy became a full professor at Stanford University in 1962, a position he held for 38 years. When he accepted his position at Stanford, McCarthy founded the Stanford Artificial Intelligence Laboratory, an institution that remains at the forefront of AI research.

McCarthy devoted much of his career to the creation of programs geared to simulate intelligence in machines. One particular field of interest to McCarthy has been the question of simulating emotions in robots. To McCarthy, emotions allow individuals to forge greater connections with others, creating a more cohesive society. However, machines by definition do not have emotions, and creating artificial versions has struck some researchers as a disquieting goal. Others have argued that simulating emotions defeats the purpose of robots, and doing so will only serve to manipulate the psyches of humans working with the machines in question.

Over the course of his career, McCarthy has made substantial advances to the development of artificial intelligence. In particular, he is recognized as the creator of the LISP programming language, the mathematical theory of computation, and the concept of computer time-sharing on supercomputer systems. McCarthy also developed computer programs that apply common-sense knowledge and reasoning, demonstrating that computer programs do not require preprogrammed responses to every logical quandary. As a result of his achievements, McCarthy has received some of the most prestigious awards in the fields of computing technology, artificial intelligence, and mathematics. In 1971, he won the Turing Award for his contributions to the field of artificial intelligence. This award, given by the Association for Computing Machinery in New York City, recognizes contributions of lasting and major technical importance to the computer field. He received the Kyoto Prize in 1988, and in 1990, he won the American National Medal of Science in the field of Mathematics and Computer Science for his lifetime achievements. This yearly award, akin to the Nobel Prize, has been handed out since 1959 and has been given to 468 recipients since its inception. In 2003, McCarthy won the Benjamin Franklin Medal in Computer and Cognitive Science, awarded by the Franklin Institute in Philadelphia.

McCarthy's philosophical arguments about the nature of intelligence, and the creation of artificial intelligent systems, remain a dominant school of thought within computer science. Although he has transitioned to emeritus status at Stanford, McCarthy remains active within the discipline and continues to produce scholarly research and contribute to debates on the feasibility and desirability of artificial intelligence, particularly in the development of military robots and autonomous weapons. The Stanford Artificial Intelligence Laboratory and the MIT Artificial Intelligence Laboratory remain among the top research locations for AI in the United States.

Sources

Hilts, Philip J. 1982. *Scientific Temperaments: Three Lives in Contemporary Science.* New York: Simon & Schuster.

McCorduck, Pamela. 2004. *Machines Who Think: A Personal Inquiry into the History and Prospects of Artificial Intelligence* (2nd ed.). Natick, MA; A. K. Peters.

Minsky, Marvin (1927–)

Marvin Minsky is both the Toshiba Professor of Media Arts and a professor of electrical engineering and computer science at the Massachusetts Institute of Technology (MIT). He is one of the cofounders of the MIT Computer Science and Artificial Intelligence Laboratory (CSAIL), and has been one of the most influential thinkers in the field of artificial intelligence for more than five decades.

Minsky was born in New York City in 1927. After a two-year stint in the U.S. Navy during World War II, he returned to civilian life and enrolled in the mathematics program at Harvard University. He received his bachelor's degree in 1950 and immediately commenced work on a doctorate in mathematics at Princeton, where he received his Ph.D. in 1954. In 1958, Minsky returned to the Boston area to join the MIT faculty, and one year later, he and John McCarthy cofounded CSAIL, creating one of the largest concentrations of artificial intelligence research in the world. In 1968, Minsky completed a tentacle robot arm, using a biomimetic concept that he believed would offer many more applications than a simple arm with limited flexibility. The engineering challenges associated with such an endeavor had long daunted many roboticists.

Minsky is perhaps best known for his theoretical work in artificial intelligence, where he has spent decades attempting to discover the fundamental nature of what constitutes thought and how it can be replicated outside of a human brain. He famously rejected the neural network concept in the 1970s, although in later years, he returned to the idea and has conceded that some aspects of it merit further examination. Minsky has argued that intelligence many simply be the result of interactions between chemicals, electrical impulses, and organic structures, which should be readily replicable under certain circumstances. Among his many accolades, Minsky received the 1969 Turing Award, the Japan Prize of 1990, and the Benjamin Franklin Medal in 1991.

Sources

Brockman, John. 2003. *The New Humanists: Science at the Edge*. New York: Barnes & Noble, 2003.

Massachusetts Institute of Technology Media Lab. 2012. *Marvin Minsky*. Retrieved from http://web.media.mit.edu/~minsky/.

Moravec, Hans (1948–)

Hans Moravec is a professor at the Carnegie Mellon University Robotics Institute who conducts research in robotics, artificial intelligence, and technology's effects on humanity and human society. He is a futurist who believes that humans are in the process of evolving into cybernetic organisms that will seamlessly integrate the most useful aspects of machinery directly into our bodies, creating a much-enhanced individual, though not a true new species, as the nonorganic enhancements will need to be added to each generation, rather than be passed down to offspring.

Moravec was born in 1948 in Austria and emigrated to Canada with his family prior to pursuing his higher education. He completed a bachelor's degree in mathematics at Acadia University in 1969, with a master's degree to follow at the University of Western Ontario in 1971. Moravec earned his doctorate at Stanford University in 1980, writing his dissertation about a unique robot that he designed and built. His robot included a television camera that broadcast images to a remote computer entirely separate from his ambulatory robot. The computer was then able to give directions to the robot, allowing it to navigate through a series of obstacle courses while making a three-dimensional map of its environment. The concept, of a computer with substantial information processing handling the cognition for a relatively simple robot pioneered the concept of remotely piloted vehicles using a central control unit and inspired many of the later conceptions of potential swarms of small robots being directed by a central processor.

In 2000, Moravec published his thoughts about the transition of humans from organic to cybernetic existence in *Robot: Mere Machine to Transcendent Mind*. In the work, he argued that robots would transcend human thought capabilities by 2040, putting the "singularity" for such an event quite a bit later than many futurists making similar predictions. He also presented the idea that Moore's Law, which was originally postulated solely in regard to integrated circuits, could be applied to many other technologies, not all of them directly related to computers. Moravec believes that robotic design will continue down a path toward complete independence from humans, in that they will be designed by fellow robots rather than human engineers. To Moravec, this would constitute the rise of the first artificial species, in that the robots would be capable of both replication and self-improvement. In 2003, Moravec founded SEEGRID Corporation, with the avowed goal of creating a truly autonomous robot capable of navigating its entire environment without any human control or input. The military implications if such a machine is completed are obvious—without the need for human input, no service personnel would be placed at risk in a combat environment. Of course, without human control, such a robot given over to military service might well signal the dystopian future that Moravec insists will not occur.

Sources

Carnegie Mellon University Robotics Institute. 2012. *Hans P. Moravec.* Retrieved from http://www.frc.ri.cmu.edu/~hpm/hpm.cv.html.

Moravec, Hans P. 2000. *Robot: Mere Machine to Transcendent Mind.* New York: Oxford University Press.

Obama, Barack (1961–)

Barack Obama is the 44th president of the United States. He was born in Hawaii in 1961. After initially enrolling at Occidental University, he completed his bachelor's degree in political science at Columbia University in 1983. He completed his law degree at Harvard University in 1991 and accepted a position teaching at the University of Chicago Law School. In 1996, he commenced his political career, winning election as a state senator in Illinois. After serving four terms as a state senator, he won election to the U.S. Senate in 2004, and in 2008, he was elected the nation's first African American president.

When Obama assumed the office of the presidency, he inherited two unpopular wars from his predecessor, President George W. Bush. During his campaign, he had suggested a desire to end combat operations in each location, although he did not expressly promise to end either conflict. Early in his presidency, he announced that all American combat troops would leave Iraq by the end of 2010, with all American forces to leave by the end of the following year. At the same time, he deployed additional personnel to Afghanistan, demonstrating that the former safe haven of Al Qaeda would once again be the primary military focus for the United States.

Like Bush, Obama's military strategy has proven highly reliant on robotic platforms, including a sharp increase in the inventory and utilization of unmanned aircraft. In many ways, UAVs are the perfect solution to reconciling his desire to maintain pressure on Al Qaeda and its affiliated groups while reducing the number of American personnel who are directly in harm's way in the combat zones. As such, it is unsurprising that the number of armed UAV strikes has increased precipitously during the Obama administration. In addition to increasing the commitment of UAVs over Iraq and Afghanistan, the Obama administration has broadened the number of locations where UAVs have been employed, with strikes by American UAVs confirmed in Libya, Pakistan, Somalia, Uganda, and Yemen, and alleged in other locations.

Overall, the use of UAVs has been a public relations boon for the administration, as they allow a demonstration of force and an erosion of Al Qaeda's leadership without increasing the strain on the military personnel system. In the months leading up to the raid upon Osama bin Laden's compound in Abbotabad, Pakistan, a number of different unmanned platforms were used to conduct surveillance of the area. At the same time, the president authorized the use of armed unmanned aircraft in attacks against the leadership of Al Qaeda in the Arabian peninsula in Yemen, Joseph Kony and the Lord's Resistance Army in Uganda, Al Shabaab in Somalia, and Moammar Gaddhafi in Libya. In each case, the use of UAVs has largely been hailed as an effective application of military power without forcing the United States to deploy ground troops to the location. Many of the UAV activities have been carried out by remote pilots operating aircraft for the Central Intelligence Agency, rather than the military, a fact that has had interesting legal ramifications.

An unexpected controversy has gradually unfolded over a single airstrike in Yemen. In April 2010, President Obama specifically authorized the targeted killing of Anwar al-Awlaki, the regional commander of Al Qaeda in the Arabian peninsula. An American citizen by birth, al-Awlaki had repeatedly called for jihad against the United States, and had been involved in the inspiration and planning of a number of terror attacks within the United States. Some legal scholars have argued that al-Awlaki should not have been targeted without due process because he was entitled to constitutional protections. Others have argued that he was actively involved in warfare against the United States and thus could not claim immunity from retaliation. The point became moot on September 30, 2011, when two MQ-1 Predators fired missiles at a vehicle containing al-Awlaki, killing him and his three companions.

President Obama's preference for UAVs and other robotic systems can be inferred through the procurement practices of the Pentagon, as well as guiding national policy documents such as the Quadrennial Defense Review, the National Security Strategy, and the National Military Strategy, each of which has placed greater emphasis on unmanned platforms during his presidency. UAVs have occasionally caused international problems for the president. The parliament of Pakistan has repeatedly called for an end to drone strikes in its country, although the strikes have continued apace against Al Qaeda militants. In one of the more embarrassing UAV incidents in history, on December 4, 2011, the government of the Islamic Republic of Iran announced the capture of an RQ-170 Sentinel, one of the most carefully guarded and highly classified UAVs in the American inventory. Details of the capture remain sketchy, but it is apparent that the aircraft belonged to the CIA and was engaged in surveillance of Iranian nuclear facilities. In April 2012, the Iranian military claimed that it had successfully decrypted the data contained within the aircraft, and that it was in the process of building a replica. American officials have argued that Iran does not have the technological capability for such a project, although China and Russia might seek to obtain information on it as a means of advancing their own programs.

Sources

Mendell, David. 2007. *Obama: From Promise to Power*. New York: HarperCollins.

Obama, Barack. 1995. *Dreams from My Father: A Story of Race and Inheritance*. New York: Three Rivers Press.

Panetta, Leon (1938–)

U.S. Secretary of Defense Leon Panetta was appointed to the position in 2011 by President Barack Obama. He previously served as the director of the Central Intelligence Agency. In both positions, Panetta has overseen substantial unmanned aerial vehicle programs and has pushed for greater reliance on such systems for intelligence, surveillance, and reconnaissance operations. He was born in 1938 in Monterey, California, to Italian immigrant parents. In 1960, Panetta completed a bachelor's degree in political science at Santa Clara University. Three years later, he finished a law degree from the same school, after which he joined the U.S. Army for a two-year stint in military intelligence. After working for a series of politicians and opening a private law practice, Panetta was elected to Congress in 1976.

Panetta remained in Congress until 1993, eventually rising to chair the House Budget Committee from 1989 to 1993. That year, President William Clinton appointed him to head the Office of Management and Budget. In 1994, he became Clinton's White House chief of Sstaff, a position he held for three years. He returned to California for a decade in academia, and in 2006, participated in the Iraq Study Group, offering recommendations on the Iraq War to President George W. Bush. In 2009, President Barack Obama nominated Panetta to head the CIA. During his confirmation hearings, he referred to UAV strikes as the most effective weapon currently in use against Al Qaeda. Upon confirmation, he increased the number of armed UAV flights being conducted by the CIA, resulting in more than 50 Al Qaeda militants killed by strikes just in the month of May 2009. In 2011, President Obama tapped Panetta to replace the retiring Robert M. Gates as the secretary of defense. In his new role, Panetta continued to emphasize the need to retain pressure on Al Qaeda through aggressive use of unmanned aircraft. His successor at the CIA, General David Petraeus, has continued Panetta's policies regarding UAVs, with the result that the two primary users of armed UAVs, the CIA and the Pentagon, are in almost complete harmony regarding their utility over the modern battlefield. Panetta has also pushed for increasing numbers of unmanned ground vehicles and naval vessels, seeing them as the technological wave of the future that offers almost limitless potential for the American military.

Sources

Clinton, Bill. 2005. *My Life*. New York: Vintage.

U.S. Department of Defense. 2012. *Leon E. Panetta*. Retrieved from http://www.defense.gov/bios/biographydetail.aspx?biographyid=310.

Reddy, Raj (1937–)

Dabbala Rajagopal "Raj" Reddy is one of the academic world's leading developers of artificial intelligence and robotics. He has worked and taught at Stanford University, Carnegie Mellon University (CMU), and in his native country, India. He was the first director of the Robotics Institute at CMU and helped to found Rajiv Gandhi University of Knowledge Technologies in 2008. The new school has four campuses in the Indian state of Andhra Pradesh, and Reddy serves as the chancellor and chairman of the university's governing council. He also chairs the governing council of the International Institute of Information Technology in Hyderabad.

Reddy was born in Andhra Pradesh and studied engineering at Guindy, graduating in 1958. Two years later, he completed a master's degree at the University of New South Wales in Australia. Before beginning his doctoral studies at Stanford University, Reddy worked for IBM's Australian branch. At Stanford, Reddy worked with artificial intelligence pioneer John McCarthy and was the first of many McCarthy students to complete a Ph.D., graduating in 1966. He remained on campus as an assistant professor for three years before accepting a tenured position at CMU in 1969.

Only seven years after defending his dissertation, Reddy received a promotion to full professor at CMU. In 1979, he cofounded the Robotics Institute, creating an organization that quickly became one of the world's largest and most-respected organizations dedicated to the research and development of robots. Reddy remained the director for its first 12 years, before accepting the position of dean of the School of Computer Science, a role he fulfilled from 1991 to 1999.

In addition to facilitating the education of future robotics pioneers, Reddy has carried out his own active research agenda for four decades. His area of interest is the portion of human intelligence dedicated to sensory perception and motor control. Reddy believes that understanding how these mechanisms function in organic systems is the key to emulating such systems in machines. As such, his research has underpinned many of the sensory input devices created at the Robotics Institute. Reddy also examines the particular problems associated with speech recognition by machines, and he worked extensively with the Defense Advanced Research Projects Agency (DARPA) effort to create a speech understanding research program.

One of Reddy's driving motivations is the idea that technology should serve society, and that it might offer the key to alleviating or eradicating world poverty. He has pushed to create digital libraries with open access to materials, believing that the proliferation of Internet access will allow educational opportunities to the masses of poor people throughout the world. His work in this area earned him the French Legion of Honor in 1984. Reddy also received the Turing Award for contributions to artificial intelligence in 1994 and the Okawa Prize for advancing information and telecommunications in 2004. In 2005, Reddy received the Honda Prize in recognition of his contributions to the fields of robotics and computer science. He won the Vannevar Bush Award in 2006 in recognition of his contributions to science and statesmanship. His efforts to bring educational opportunities to rural youth in his native India continue to expand, and India is emerging as a world leader in the development of robotic systems, in no small part due to his efforts.

Sources

Carnegie Mellon University Robotics Institute. 2012. *Raj Reddy.* Retrieved from http://www.rr.cs.cmu.edu/.

Nilsson, Nils J. 1994. *Dabbala Rajagopal "Raj" Reddy.* Retrieved from http://amturing.acm.org/award_winners/reddy_6247682.cfm.

Simon, Herbert A. (1916–2001)

Herbert Simon was an American polymath who made major contributions to the fields of political science, economics, sociology, and psychology, with substantial influence over the fields of artificial intelligence and computer science. He was born in Milwaukee, Wisconsin, in 1916, and earned both his bachelor's degree (1936) and doctorate (1943) at the University of Chicago, both in the field of political science. After short stints at the University of California–Berkeley, and the Illinois Institute of Technology, Simon accepted a position at the Carnegie Institution of Technology (later Carnegie Mellon University [CMU]) in 1949. He remained in this position for the rest of his life and continued to teach in multiple academic departments, including computer science, business, philosophy, social and decision sciences, and psychology, until his death in 2001. While at CMU, he interacted with many robotics pioneers, in part due to his own investigations regarding the nature of artificial intelligence.

Simon initially focused on the nature of human decision making from a social scientist's perspective, but his intellect simply refused to be bound within the confines of a single discipline. As he continued research into the fundamental concepts of intelligence, knowledge, and memory, he began to work with individuals that wished to replicate these ideas in an artificial environment. In 1956, Simon, Allen Newell, and J. C. Shaw completed their Logic Theory Machine, a computer program designed to imitate the problem-solving process of a human brain. The program has been called the first artificial intelligence ever designed, although the term itself was not coined until the following year. It included a search tree process that included a series of heuristics to prevent the program from following logic paths that were unlikely to bear any fruit. Although confined to proving mathematical theorems, and hence not tackling every aspect of human intelligence, it represented a major leap forward in the design of machine logic systems. Not only did it successfully prove a series of complex mathematical theorems, it actually created several new proofs considered more elegant than the classical approach.

Simon considered "artificial intelligence" to be a concept that encompassed two entirely separate forms of thought processing.

"Cold cognition" includes the use of reason, logic, perception, and decision making. It is relatively easy to reduce the rules of cold cognition into mathematical representations, and hence, into a manner that machines can understand. "Hot cognition," in comparison, involves emotions, pain, pleasure, and desires, and is both unpredictable and not replicable from one human being to the next. While reason and logic might be shared, emotional reactions to shared stimuli are never precisely the same, and hence, may be beyond the ability of a machine's experience. Simon recognized that in some aspects of knowledge and intelligence, computers have an inherent advantage. His theory of human expertise, for example, posits that a human requires approximately 10 years in a field in order to acquire sufficient knowledge to qualify as an expert. Simon argued that expertise requires the acquisition of approximately 50,000 discrete pieces of information, and humans have a finite limit to how much new information they can incorporate. In comparison, machines do not have the same limitations on processing speed or fatigue, and can essentially become instant experts in almost any field within the cold cognition framework, while humans very rapidly become experts in hot cognition, through instinct, and experience. Over the course of his career, Simon amassed an enormous publication record of more than 1,000 works, making him one of the most-cited American academic figures of the 20th century.

Sources

Crowther-Heyck, Hunter. 2005. *Herbert A. Simon: The Bounds of Reason in America*. Baltimore: Johns Hopkins University Press.

Simon, Herbert A. 1991. *Models of My Life*. New York: Basic Books, 1991.

Turing, Alan (1912–1954)

Alan Turing was born in London on June 23, 1912. He was identified as a mathematical genius at an early age and worked during World War II (1939–1945) as one of the United Kingdom's top cryptographers at Bletchley Park. He is often referred to as the father of both computer science and artificial intelligence, as he substantially contributed to each field in the last decade of his life.

He provided many of the foundational definitions of computer science and is best known for the Turing Test, a means of determining if a machine should be considered to possess intelligence.

Turing attended King's College–Cambridge, to study mathematics, graduating with honors in 1934. He then spent a year as a fellow at King's College before enrolling in a doctoral program at Princeton University. At Princeton, Turing worked under the tutelage of Alonzo Church and earned his Ph.D. in under three years, graduating in June 1938. He subsequently returned to King's College, and simultaneously began to work with the British Government Code and Cypher School. This work required him to maintain a top security clearance, a fact that later came into conflict with his private life and ultimately led to his demise.

During World War II, Turing became a key cryptanalyst at Bletchley Park, working diligently to crack German ciphers, in particular those created with the Enigma machine. Although most of Turing's work remained classified for decades, many of his fellow cryptographers considered Turing the most effective individual at Bletchley Park. He travelled to the United States during the war to accelerate American efforts to decipher German and Japanese naval codes, renewing some of the relationships that he had formed at Princeton. In 1945, Turing was awarded the Order of the British Empire for his wartime services.

After the war, Turing accepted a position at the National Physical Laboratory, where he assisted in the design and construction of the world's first stored-program computer, the Pilot Automatic Computing Engine. In 1948, he returned to academia, accepting a position at the University of Manchester, where he soon became the deputy director of the new computing laboratory. In addition to designing some of the earliest software for stored-program computers, Turing devoted substantial time to theoretical issues that would not become possible for decades. In particular, he studied various philosophical conceptions of the human intellect, and how it might be possible to determine if a machine possessed intelligence rather than merely a store of data and computational power.

In 1950, Turing proposed a simple test to determine if a machine could think. He suggested that a human interrogator should question both the machine and a fellow operator through a neutral medium, such as a series of teletype messages. If the interrogator could not distinguish the machine's answers from those of the human, the machine could be considered intelligent. Turing proposed in the same paper that a simpler form of artificial intelligence would be to create a machine capable of learning, which could then be subjected to a formal education in the same manner as a human child, though presumably at a faster rate.

Turing worked on cutting-edge mathematical theories in the same period that he contributed to the burgeoning field of computer science. He was fascinated by the relationship between mathematics and biology, finding a series of unexplained patterns within plant structures. Many of his most important contributions were written in the 1950s but remained unpublished for four decades. When finally rediscovered and released, they became dominant works within the field of mathematical biology, demonstrating how far advanced his ideas were for their time.

Turing's private life, and the existing British indecency laws of the 1950s, destroyed his career and ultimately his existence. In 1952, Turing was charged with gross indecency for engaging in homosexual activities. When he pled guilty to the charge, he was offered the choice of chemical castration or imprisonment. Hoping to continue his academic career, Turing chose to receive estrogen injections. At the time, it was believed that homosexuals were prone to exploitation by Soviet intelligence agencies; thus, his conviction triggered a revocation of Turing's security clearance. In turn, this ended his consulting work on cryptography for both the British and American governments. There was no evidence that Turing had ever considered espionage, or even been approached by a foreign intelligence agent, but he was still regarded with suspicion and placed under surveillance by the British government.

On June 8, 1954, Turing's housekeeper arrived at his home and found his body on the floor of his bedroom. Near the corpse was a half-eaten apple. A medical examiner ruled the cause of death to be cyanide poisoning, with the deadly agent most likely delivered through the apple, which was not retained for chemical analysis. While an inquest determined that Turing had committed suicide, his mother and others argued that his death was accidental and due to carelessness in his home laboratory.

Many of Turing's achievements were not recognized until long after his death. However, his influence on the development of computer science is unparalleled. In 1966, the Association for

Computing Machinery instituted the annual Turing Award, given to an individual for contributions to the computing community. Prominent recipients who have contributed to the development of robotics include Marvin Minsky (1969), John McCarthy (1971), Dabbala Rajagopal "Raj" Reddy (1994), Frances Allen (2006), Barbara Liskov (2008), and Leslie G. Valiant (2010). In 2009, British Prime Minister Gordon Brown formally apologized on behalf of the government for the way Turing had been treated near the end of his life.

Sources

Hodges, Andrew. 1983. *Alan Turing: The Enigma*. New York: Simon & Schuster.

Leavitt, David. 2006. *The Man Who Knew Too Much: Alan Turing and the Invention of the Computer*. New York: W. W. Norton.

Teuscher, Christof. 2004. *Alan Turing: Life and Legacy of a Great Thinker*. New York: Springer, 2004.

Vaucanson, Jacques de (1709–1782)

Jacques de Vaucanson was an 18th-century French inventor who built a series of automata for the entertainment of wealthy and noble patrons throughout Europe. He also designed and constructed a number of industrial innovations, including an automated loom that accepted patterns from punch cards, a forerunner to modern industrial production. While de Vaucanson did not construct true robots, many of his machines inspired successors to continue experimenting in the early development of autonomous machines.

de Vaucanson was born and raised in Grenoble, and came from a large, poor family. He was fascinated by the precision of clocks and initially intended to become a clockmaker. He also demonstrated an interest in organic systems, spending a short time as a surgeon's apprentice before beginning work as a machine inventor. In 1727, Vaucanson began experimenting with humanoid automata, completing his first model, The Flute Player, in 1737. This life-sized figure, with fingers covered in skin, could play a dozen songs on a simple pipe. Within a year, he completed The Tambourine Player, continuing his theme of musical automata for the amusement of his patrons.

de Vaucanson's most famous work was The Digesting Duck, which was completed in 1738. This duck, which Vaucanson claimed could simulate all of the activities of a live specimen except flight, could flap its wings through a detailed mechanism containing more than 400 parts. It could eat, drink water, and defecate, indicating the possession of an entire digestive tract. In reality, the "eaten" grains went into a stomach pouch, and a paste of "digested food" passed from a hidden storage compartment through rubber tubing, with no actual digestion occurring. At the time, these automata swept the imagination of Europe's upper class, and de Vaucanson's duck became the talk of royal courts across the continent. Unfortunately, none of his inventions survived into the 20th century, nor did his design specifications. However, the duck's intricacy required substantial innovation on de Vaucanson's part, including production of the first flexible rubber tubing.

de Vaucanson shifted his attention to industrial machines in the 1740s, developing a number of new devices in addition to his automatic loom. While his inventions triggered an invitation to join the prestigious Académie des Sciences, they also provoked substantial derision from the working class. In particular, artisans such as weavers, who felt that de Vaucanson's inventions threatened their livelihood, attacked him in the streets of Paris. They found his attempts to alleviate their toil to be an abomination, and they abhorred reducing the artistry of their craft to punch-card mass production. Nearly 200 years later, de Vaucanson's punch-card system remained a primary mechanism to input data into early computers, keeping his efforts relevant well into the 20th century.

Sources

Riskin, Jessica. 2003. "The Defecating Duck, or, the Ambiguous Origins of Artificial Life." *Critical Inquiry* 29, no. 4: 599–633.

Singer, Peter W. 2009. *Wired for War*. New York: Penguin.

Warwick, Kevin (1954–)

Kevin Warwick is a professor of cybernetics at the University of Reading in the United Kingdom, and one of the dominant members of the robotics academic community who also studies artificial intelligence and biomedical engineering. He was born in 1954 in Coventry and left school at age 16 to join British Telecom. He soon chose to return to formal schooling, receiving a bachelor's degree at Aston University in 1976 and a Ph.D. from Imperial College–London. Upon completion of his doctorate, Warwick taught at the University of Oxford, Newcastle University, and the University of Warwick before joining the Reading faculty. He is currently working on a project to use cultured neural networks to control robots, in essence, growing an organic computer to control a machine.

Warwick is one of the few roboticists who sees humanity and robots becoming increasingly intertwined, rather than separated. While many futurists project a vision of robots and humans eventually coming into conflict, or robots simply surpassing their organic counterparts in virtually every fashion, Warwick believes that cybernetic organisms, combining the best elements of both, are the wave of the future. He argues that humans have already begun the process through artificial enhancements, such as artificial limbs and cochlear implants, and that the process will continue unabated. To Warwick, there is no reason to remain a "mere" human when so many potential enhanced capabilities will be available in the near future. He has also argued that such enhancements might begin in military forces and be provided on a temporary basis, to improve service capabilities, but the additions might also become a source of conflict if they are available only to the economically advantaged.

Warwick has been heavily involved in debates within the discipline of robotic ethics, which complements his interest in the ethics of human enhancements. He studies the self-organization behaviors of robots that have not been programmed to behave in a certain fashion and finds that the less preprogramming is forced into a robotic control system, the most likely a surprising outcome will be the result. His viewpoint is one of the underpinnings of the school of robotic design that calls for swarm behavior in simple robots. Given his desire to allow ambiguous responses to external stimuli, it is unsurprising that many of his designed robots have shown unexpected and unpredictable behavior when they encounter new environments. One of the best-known examples is detailed in Warwick's autobiography, *I, Cyborg*, in which a robot could not reconcile its programming with its situation and essentially committed suicide in response. Warwick also advocates the idea of robots teaching one another, without any human intervention in the transfer of data, believing it will create a far better result in robotic design.

Warwick is probably best known for using himself as an experimental subject. In Project Cyborg, Warwick had a series of computer chips implanted into his body to test some of the basic assumptions about the possibility of cybernetics in humans. In the first stage, he had an radio-frequency identification (RFID) transmitter implanted, with the purpose of controlling automatic doors, lights, and other appliances based on his proximity. Next, he had an electrode array tapped directly into his radial nerve to create a neural interface. With a bit of experimentation, his team was able to isolate the frequency and strength of human electrical impulses, and soon he was able to control a robotic arm using only his mental impulses. Force feedback sensors on the robotic arm allowed his brain to perceive the sensation of the robot arm touching items in its environment. Warwick's wife agreed to have a simpler array implanted, and the two were able to engage in basic communication with one another using only nervous system electrical impulses, essentially engaging in a form of telepathy. Interestingly, not only did Warwick's body not reject the electrode array, his nerve tissue began to grow around the electrodes, essentially incorporating them into his body's normal function.

Warwick sees Project Cyborg as a necessary first step in the creation of new medical devices to help individuals with damaged neural systems. While such inventions might be the subject of future research, the first stage of the project has already triggered the company VeriChip to offer an implantable tracking device that can be connected to a global positioning system (GPS), thus theoretically allowing individuals at high risk of kidnapping to be rapidly found. For his achievements, Warwick has been given the Future of Health Technology Award from the Massachusetts Institute of Technology, the University of Malta medal, the Institution of Electrical Engineers Senior Achievement Medal, the Mountbatten Medal, and made an Honorary Member of the Academy of Sciences in St. Petersburg.

Sources

Warwick, Kevin. 2012. *Professor Kevin Warwick*. http://www.kevinwarwick.com/.

Warwick, Kevin. 2002. *I, Cyborg*. London: Century.

Whittaker, William "Red" (1948–)

William "Red" Whittaker is the Fredkin Research Professor at Carnegie Mellon University's Robotics Institute in Pittsburgh, Pennsylvania. He has been a dominant force in academic robotics research for more than three decades, and has been instrumental in the design and production of autonomous vehicles. Most of his work has centered around the production of practical designs intended to solve specific real-world problems rather than demonstration models to prove scientific concepts.

Whittaker was born in Hollidaysburg, Pennsylvania, in 1948. His father, an explosives salesman, and his mother, a chemist, stimulated his interest in science and engineering. After serving in the U.S. Marine Corps, he earned a bachelor's degree in civil engineering at Princeton University, graduating in 1973. He completed his graduate work in the same field at Carnegie Mellon, earning his Ph.D. in 1979.

Whittaker's first field robotics experience came in 1979, when he led a team that constructed robots to inspect and repair damaged areas in the nuclear reactor at Three Mile Island. This led to a succession of robotic models designed for extremely hostile environments, including the Chernobyl reactors, the craters of active volcanoes, the icy landscape of Antarctica, and eventually the surface of Mars. To facilitate this work, Whittaker cofounded RedZone Robotics in 1987.

Recently, Whittaker has become interested in the design and production of autonomous vehicles. In part, he argues that such robots will actually reduce the number of vehicular fatalities in the world, as they will be capable of much quicker reactions to changing conditions and are not subject to the distractions that often impede human drivers. His teams have performed extremely well in the DARPA Grand Challenge competitions, beginning with the first contest in 2004, when his robot, Sandstorm, travelled farther on the course than any other entrant. The following year, Whittaker's teams took second and third place, bested only by a former CMU collaborator, Sebastian Thrun of Stanford University's Artificial Intelligence Laboratory. In 2007, Whittaker's team won the $2 million prize when their entry, Boss, autonomously navigated a 60-mile urban course. In this iteration of the Grand Challenge, the vehicle was required to obey all traffic laws and interact with other vehicles on the roadway, a substantially greater difficulty than the empty desert navigation of previous Grand Challenges.

Whittaker remains on the cutting edge of robotic design, and many of his concepts have direct or indirect military applications. His current projects include further development of robots capable of operating in environments too remote or dangerous for humans as well as machines designed to perform laborious tasks such as grain harvesting or excavation without human oversight. His designs also include robots that interact directly with humans, such as RoboHost, a robotic tour guide for museums. He is a university professor at CMU, where he also serves as the director of the Field Robotics Center. He has received the Engelberger Technology Award, the Design News Special Achievement Award, the Hero of Manufacturing Award, and the Aviation & Space Technology Award. In addition to his own work, Whittaker has advised more than two dozen doctoral projects, making him one of the most influential roboticists in academia today.

Sources

Arndt, Michael. 2006. " 'Red' " Whittaker: A Man and His Robots." *BusinessWeek*. June 25. http://www.businessweek.com/stories/2006-06-25/red-whittaker-a-man-and-his-robots.

Whittaker, William L. 2012. *William L. "Red" Whittaker*. http://www.frc.ri.cmu.edu/users/red/.

Wilson, Daniel (1978–)

Daniel Wilson is a futurist writer with the solid academic background necessary to understand the likelihood of his own predictions, and as such, has become one of the most well-known popular writers on the ramifications of robotics, bridging the gap between the theoretical work of robotics pioneers and the public's grasp of what is both possible and likely. He was born in Tulsa, Oklahoma, in 1978 and earned his bachelor's in computer science at the University of Tulsa in 2000. He holds master's degrees in robotics and machine learning from Carnegie Mellon University, where he also earned his doctorate in 2005.

Wilson's dissertation project studied the use of automatic sensors for location and activity monitoring of people within a structure, with potential applications in the home consumer market.

Wilson's short work career has included stints at Microsoft, Xerox, Northrop Grumman, and Intel Corporation, but his most prominent role has been through his frequent publications, both fiction and nonfiction, in which he discusses the future interaction of humans and robots. His tongue-in-cheek *How to Survive a Robot Uprising* proved extremely popular upon its release in 2005, and his subsequent works have also received both critical praise and commercial success. Wilson is at the forefront of a group of writers working to raise the public's comfort level with the notion of a robotic revolution, particularly through his writings in *Popular Mechanics* and other similar well-known magazines and journals.

Sources

Deahl, Rachel. 2010. "Daniel H. Wilson: A Hollywood Favorite Awaits His Publishing Moment." *Publisher's Weekly* 257, no. 50. December 20. http://www.publishersweekly.com/pw/by-topic/book-news/page-to -screen/article/45572-daniel-h-wilson-a-hollywood-favorite-awaits-his -publishing-moment.html.

Wilson, Daniel H. 2012. *Daniel H. Wilson*. Retrieved from http://www .danielhwilson.com/.

6

Data and Documents

Military robots have already begun to deploy with the human forces engaged in conflicts around the globe. These robots come in a wide variety of sizes, functions, and sophistication, but all serve to advance the military prowess of the nations that employ them, while in most cases keeping human personnel further from harm's way. They serve as intelligence-gathering platforms, cargo transport devices, point defense weapons, and machines that directly attack the enemy. This chapter is designed to demonstrate the variety of systems that have already seen service on the modern battlefield or are expected to do so in the extremely near future. The information is derived entirely from open sources. Individuals interested in examining certain robots in greater detail are strongly urged to begin their quest in two locations. First, most of the manufacturers of military robots make significant information about their devices readily available online, in large part because they are hoping to sell copies of the systems. Second, digital videos of most of the systems are posted to websites such as YouTube on a regular basis and can provide an interesting look at the capabilities of all but the most classified systems. At the end of the chapter, there are excerpts from a number of key documents that have influenced the decisions to create and utilize robotic weapon platforms on the modern battlefield.

BigDog

For decades, fiction writers and filmmakers have envisioned robotic versions of organisms, while roboticists have focused on wheeled and tracked models. Some have argued that there is no compelling reason for machines, designed by humans, to emulate organisms, the product of millions of years of environmental adaptation and evolution. While no organisms have evolved locomotion involving wheels or tracks, both mobility approaches have proven far easier to engineer in robots. Mimicking the movements of legs, whether based on insects, reptiles, or mammals, has provided a difficult challenge, as biological legs incorporate many muscles capable of extremely fine motor control.

The desire for legged robots is due to the inherent flexibility of legs—a legged robot might be able to traverse all of the terrain that humans can cross but that is inaccessible to vehicles. Better yet, a legged robot might be able to travel as an insect does, virtually ignoring the effects of slopes and slippery surfaces. For decades, the largest design problems were derived from inaccurately mimicking the organisms being studied. Robotics engineers either tried to manage all of the legs from a central system, or they moved all of the legs on one side of the robot in unison, neither of which emulates the function of biological systems. When they began to experiment with independently controlled legs, none of which required direct control from a central processor, the robots began to function much better and faster, and to demonstrate the potential of this approach (Raibert et al. 2008, 1–2).

A robot that can negotiate the same terrain as an insect remains a major challenge. However, Boston Dynamics, working in conjunction with Foster-Miller, Harvard University, and the National Aeronautics and Space Administration (NASA) Jet Propulsion Laboratory, has unveiled BigDog, a quadruped robot designed to move wherever infantry forces may go, while carrying a substantial payload in the process. BigDog represents the 21st-century version of the Army mule. Like many far-reaching projects, the BigDog development process began with funding from the Defense Advanced Research Projects Agency (DARPA), which has now turned responsibility for further research and enhancement over to the military and private firms (Raibert et al. 2008, 2).

The aptly named robot stands approximately 30 inches tall and is three feet long. It weighs 240 pounds but can carry up to 340 pounds of cargo at five miles per hour, making it able to keep up with all but the most rapid of dismounted infantry maneuvers. Testing has demonstrated that BigDog can move up inclines of 35 degrees, a slant that often presents difficulties to a human, particularly one laden with a substantial amount of equipment. BigDog has little difficulty moving across uneven ground and has proven capable of navigating icy terrain. In fact, its platform is so stable that it remains upright even when kicked from the side, showing that it can compensate for rapid changes in the operating environment without toppling over.

Each leg contains four actuators, including two at the hip, one at the knee, and one at the ankle. Each actuator uses hydraulics driven by a central pump, which in turn is powered by a simple two-stroke, 15-horsepower gasoline engine. Because each leg is driven individually, the overall system actually has substantially more stability and versatility than a human. Each leg has a variety of sensors to determine its position and the pressure required to remain steady. The robot also has a stereovision and laser gyroscope, creating a redundant ability to maintain balance. By allowing each leg to operate independently while drawing information from a single sensory system, the robot does not overload a single processor, and the response time to the individual conditions faced by each leg is greatly reduced (Singer 2009, 92).

While BigDog is still in development, it is likely to be deployed on at least a test basis in the near future. In an operating environment such as Iraq, there is little demand for a robotic mule, as more traditional nonrobotic systems can fulfill the resupply needs of units in combat with little difficulty. In the rough terrain of Afghanistan, where search and sweep operations face challenging, mountainous terrain, BigDog may provide a vital edge for dismounted patrols.

Dragon Eye

The RQ-14 Dragon Eye is a small tactical unmanned aerial vehicle (UAV), designed specifically for the U.S. Marine Corps. It was created by the combined efforts of researchers at the Naval Research Laboratory and the Marine Corps Warfighting Laboratory. Once it was deemed ready for full-scale production, the contract to build the system went to AeroVironment, also the producer of the RQ-11 Raven. Over 1,000 Dragon Eyes were produced, with most designated for service in Iraq.

The Dragon Eye is only four feet wide and three feet long, and its launching weight is less than six pounds. It is powered by an electric motor that can run for 60 minutes before needing a recharge. The Dragon Eye has an extremely limited ceiling and is not designed to fly higher than 500 feet. This limitation has little practical effect, as the system is used primarily for battlefield scouting over urban areas. The UAV is extremely quiet, allowing the operator to identify targets without alerting them to the small aircraft's presence. It is launched either by hand or via a small bungee catapult and can navigate using both global positioning satellites (GPS) and inertial navigation systems (AeroVironment, 2011).

The Dragon Eye has a pair of relatively unique attributes that differentiate it from similarly sized UAVs designed for tactical employment. First, while it is flown by a single operator using a modified laptop computer, the user also wears a pair of goggles with an internal camera display, essentially allowing him or her to see what the Dragon Eye sees. Second, the unit is designed to break apart upon impact with the ground or any other obstacle. By creating natural fault lines, much of the kinetic energy from a crash, which would tend to damage or destroy similar vehicles, is dissipated. Because the breakaway components can be reassembled or replaced as necessary, system maintenance is relatively quick and simple.

Dragon Runner

Dragon Runner is a small tactical ground robot designed to provide intelligence, surveillance, and reconnaissance (ISR) to small units of dismounted troops. It is easily portable, weighing less than 10 pounds, and field tests have proven it to be extremely durable. It is also very user friendly, with the user directing the robot through a simple videogame-style handheld controller (Singer 2009, 68). The Dragon Runner measures only 15 inches long, 10 inches wide, and five inches high, including its four oversized wheels. It has no required orientation; thus, it can be thrown through windows or over fences without fear that it will land upside down. It is designed to withstand an impact onto concrete from more than 20 feet and can be thrown from a vehicle moving up to 45 miles per hour (Singer 2009, 145).

The Dragon Runner is strictly an observation unit, and there are no plans to create an armed variant. While its camera resolution is not as great as that of larger machines, it is capable of limited autonomous activity. When placed in sentry mode, it uses audio and visual detectors to sense motion, alerting the operator if anything is sensed in the immediate area. While it is normally outfitted with wheels for superior mobility and maneuverability, it also has snap-on treads and flippers that can be positioned without tools in a matter of minutes. These additions reduce the unit's speed but allow it to function on a greater variety of terrain, including stairs.

The Dragon Runner exemplifies the cooperation that is common between the military, industry, and academia. It was designed at the Robotics Institute of Carnegie Mellon University, drawing funds from the U.S. Marine Corps Warfighting Lab. The system is licensed for production by QinetiQ, the same conglomerate that recently acquired Foster-Miller. The initial unit costs were approximately $50,000, but increased production of the Dragon Runner system allowed more efficiency, reducing the already low price tag even further (Crane 2005).

Fire Fly/Lightning Bug

The Fire Fly, later renamed the Lightning Bug, was a reconnaissance drone produced by Ryan Aeronautical Company (later Teledyne-Ryan). It was a modified version of the Ryan Firebee, an air-launched target drone designed for use in training ground-based air defense crews on the proper operation of their systems against inbound jet aircraft.

The Fire Fly's guidance system was exceedingly simple. It included a timer, a compass, and an altimeter. Essentially, the drone could fly a straight course for a predetermined length of time, then turn around and return along the same path. Because it was launched from underneath a DC-130 Hercules aircraft already flying at a cruising altitude of 55,000 feet, the new recon-naissance drone required no further input. A special model, soon called the Lightning Bug, was given an impressive array of electronic signal detection gear and an active radar system. This drone could then fly over Soviet air defense systems, enticing them to fire and give away the signal frequencies used to control their missiles, allowing the creation of effective jamming systems (Wagner 1982, 46–48).

For the duration of the Vietnam War, Ryan engineers continued to improve the Lightning Bug. One late 1960s model flew equipped with a chaff dispenser, dropping strips of foil to confuse enemy radars. Years later, it was used to rain propaganda leaflets down on enemy civilian populations. In 1972, a variant with a television camera began to broadcast images to a DC-130 Hercules controller aircraft kept a safe distance away. Several of these variants overflew the North Vietnamese prisoner of war compound known as the Hanoi Hilton in 1972 and 1973, confirming the presence of captured American personnel (Wagner 1982, 201–206; Wagner and Sloan 1992, 11). The danger of Lightning Bug missions meant most flew only a few times before being lost to enemy action, but the record-holder, nicknamed Tom Cat by its maintenance crew, managed 68 missions (Wagner 1982, 200; Wagner and Sloan 1992, 6). Specially modified Fire Fly drones last saw combat service in 2003, when several were used as chaff dispensers on the first nights of Operation Iraqi Freedom. They were simply programmed to fly and drop chaff until they ran out of fuel, in the process creating a corridor through which manned aircraft could fly, relatively safe from any radar-based systems.

Fire Scout

After nearly 15 years of using RQ-2 Pioneers, the U.S. Navy decided that it needed a new ISR platform that could take off and land vertically, carry a minimum of 200 pounds of cargo, loiter at a range of at least 125 miles for no less than three hours, and land on a ship in relatively heavy winds. Northrop Grumman won the Navy's ensuing design competition, presenting the RQ-8 Fire Scout unmanned helicopter as a craft that could exceed every Navy requirement. The company first demonstrated its prototype's capabilities in January 2000.

Despite its early affinity for the program, which was heightened by the ability to carry and launch the vehicle from a HumVee, making it an interesting design for the U.S. Marine Corps, the Navy reversed its position and cut funding for the Fire Scout in December 2001.

A smaller and less-experienced corporation might have collapsed with the demise of the Navy contract, or at the very least, abandoned the project as wasted development funds. Northrop Grumman, one of the largest defense contractors in the world, instead expanded the characteristics of the Fire Scout, adding the ability to carry a variety of weapons and inventing a cargo pod system that allowed the unmanned helicopter to deliver up to 800 pounds of supplies. This upgraded Fire Scout, designated the MQ-8B, immediately caught the attention of the U.S. Army, which believed that the machine could undertake close air support and tactical ISR missions. By late 2003, the Fire Scout became an Army development project, and the airframe design, which began as a modified Schweizer 330 light helicopter, soon began to resemble a scaled-down Apache attack helicopter. The addition of stub wings with weapons hard points completed the transition. The MQ-8B can be outfitted with AGM-114 Hellfire missiles, Stinger air-to-air missiles, Viper Strike glide bombs, or a new laser-guided 2.75-inch rocket system. Weapons experiments also showed that the platform is stable enough in flight and has such accurate sensors that it can serve in a sniper capacity, mounting a Barrett .50 rifle and striking human-sized targets more than one mile away.

While the Army concentrated on the multirole version of the Fire Scout, acquisitions specialists within the Navy began to take a renewed interest in the system. In 2005, more than three years after cancelling support for the RQ-8 variant, the Navy contracted to buy several MQ-8Bs for evaluation purposes. Ironically, the renewed Navy interest coincided with declining interest from the Army, which in January 2010 ended acquisition of the Fire Scout in favor of the fixed-wing RQ-7 Shadow. The Navy envisions the Fire Scout as a key aspect of its new Littoral Combat Ship (LCS) concept. Reflecting the need for small vessels capable of projecting overland power, the Navy has gradually transitioned away from aircraft carriers, although the carrier battle group remains the mainstay of current fleet organization. The Fire Scout will serve as both an ISR platform and an attack craft, and its ability to operate from a relatively small platform makes it perfectly suited for LCS coastal operations (Northrop Grumman 2010).

The current configuration of the Fire Scout is a 24-foot fuselage with a 28-foot rotor. The RQ-8 mounts three rotors, as its sensor payload is relatively light and this style reduces fuel consumption, consequently increasing the endurance of the craft to over eight hours on-station. The MQ-8B has four rotors, improving lift and adding payload capacity, but correspondingly reducing endurance (Northrop Grumman 2010). Perhaps the greatest selling point of the craft comes from its use of common components and standard NATO fuel, making maintenance and supply issues much easier than those of many competitors. In 2009, the Fire Scout deployed to the eastern Pacific on a counternarcotics trafficking cruise aboard the frigate *USS McInerney*. Fire Scouts have also been sent on ISR missions over Libya, where one was lost in June 2011, most likely due to enemy fire (Stewart 2011).

Gladiator

While the U.S. Army has largely selected relative newcomers iRobot and Foster-Miller as its key suppliers of ground robots, the U.S. Marine Corps has utilized a more traditional defense contractor, Lockheed Martin, to design and build its unmanned ground vehicles (UGV). The most advanced, the Gladiator, is a tracked, low-slung vehicle resembling a small tank. Weighing nearly a ton, it is substantially larger and more heavily armored than its closest competitor, the Foster-Miller TALON SWORDS. Gladiator was developed by the Robotics Institute of Carnegie Mellon University, which worked closely with Lockheed Martin engineers to bring the full system into production (Tiron 2004).

The Gladiator is a modular weapons system that can be fitted with a variety of armaments to suit the needs of the accompanying troops. It is designed to provide support to dismounted infantry units and has the advantage of a strong resistance to small arms fire, allowing it to maneuver in the open to bring extremely accurate fire against enemy forces. The Gladiator has an extensive sensor package and is capable of operations in all types of weather during both day and night. One module available for the system is designed to allow the robot to clear fields of antipersonnel mines.

The Gladiator is much larger and heavier than similar Army machines, reflecting a different design concept for the Marines. Whereas the smaller robots are typically extremely specialized, the Gladiator is a multipurpose robot capable of performing many missions with minimal adjustments. As such, the Gladiator is more likely to be prepared for a mission almost immediately, allowing rapid deployment of the system anywhere Marines call for support. Given the recent utility of Marines as the American quick-reaction force for immediate interventions around the globe, this functionality provides a substantial difference from the Army's increasingly light and nimble formations.

Global Hawk

The Northrop Grumman RQ-4 Global Hawk is one of the largest and most sophisticated UAVs in the world. It is the premiere unmanned ISR platform in American military service and is taking an ever-increasing share of missions previously performed by the manned U-2 Dragon Lady. The system can cruise at more than 400 miles per hour and can loiter up to 36 hours. The range of the RQ-4A is nearly 16,000 miles, with a maximum altitude of 65,000 feet. Global Hawk is one of the largest UAVs ever flown. Its initial wingspan of 116 feet grew to 131 feet in the RQ-4B model. Its current configuration is nearly 48 feet long and over 15 feet high. Empty, it weighs more than four tons, but the addition of fuel and sensors pushes the normal takeoff weight to nearly 23,000 pounds. This aircraft's payload capacity and long loiter time have led to tests of the system as a possible autonomous tanker, a role that might relieve some of the pressure to obtain a new manned tanker system to replace the aging fleet of KC-135s currently in service (Air Combat Command, 2009).

The Defense Advanced Research Projects Agency (DARPA) sponsored the initial design of the Global Hawk, maintaining control of the program until an advanced technology demonstrator was built, then transferring its further development to Northrop Grumman. The aircraft was intended as a revolutionary leap forward in ISR capabilities by UAVs. By flying high and fast, Global Hawk is essentially immune from all but the best modern air defenses and interceptors, allowing it to be flown with impunity over the most common conflict zones where the United States has recently taken an interest, including the deployment of forces. The Global Hawk's operational parameters make it safe to operate in many locations where the RQ-1 Predator and RQ-9 Reaper systems would be vulnerable to enemy action.

The Global Hawk offered such a potential increase in ISR capabilities that even the prototype models were rushed into service in Afghanistan and Iraq. Such a move would be unconscionable with a manned aircraft under anything but the most dire of circumstances, including a threat to national survival that simply did not exist in either of these theaters. However, moving an unmanned prototype to these zones was considered almost routine for a UAV system. While this allowed the system to be tested in real-world conditions, it also contributed additional stress to the already difficult problem of evaluating and improving a prototype aircraft. Three of the prototypes crashed during these operations, leading some detractors to call for the cancellation of the program.

In many ways, the Global Hawk performs the same function as a reconnaissance satellite but at a fraction of the cost and with greater flexibility. Whereas maneuvering a satellite is a slow, costly process, the Global Hawk's path can be altered in real time by a ground controller. A satellite may only rarely pass over an area, and may be overhead for a limited time, while Global Hawk can be directed to remain in a holding pattern indefinitely, subject only to the limits of its fuel supply. Its greatest value lies in the sensor payloads carried by the aircraft. The sensors are provided by subcontractor Raytheon and include a synthetic-aperture radar (SAR) as well as electro-optical (EO) and infrared (IR) cameras. Unlike most smaller UAVs, which can operate only a single sensory system at a time due to payload limitations or the need to share lens equipment, the Global Hawk can utilize its SAR and one of the cameras simultaneously. This means the SAR, which can penetrate clouds and sandstorms, can be coupled with a camera to produce useful images to commanders on the ground without the need of a technician to interpret the data.

Global Hawk's SAR has an integrated moving target indicator (MTI) that can detect vehicle-sized moving targets at ranges in excess of 60 miles away. It can then automatically notify the operator, either by sound effect or text message. The SAR can scan in several different modes, including wide-area, strip, and spot modes. The wide-area scan provides maximum area coverage, searching for sources of movement over thousands of square miles of territory in a single mission. The strip mode can survey a path 23 miles wide, transmitting images at 20-foot resolution, meaning a 20-square foot area fills a single pixel of the digital image. The spot mode examines less than four square miles at a time, but does so with a six-foot resolution, allowing the detection of individual people on the ground. The Global Hawk is not limited to visual sensors. It can also be fitted with extra payloads designed for electronic signal interception, with the intercepted data then immediately relayed over the same data links.

The Global Hawk provides its own navigation, using GPS technology and internal inertial navigation. This allows the aircraft to fly completely autonomously, sending images to ground stations via direct downlink, for stations within line-of-sight, and satellite downlink for other recipients. The data can be sent at speeds of up to 50 megabits per second, but enhancements may be required in the future as more sensitive collection systems, such as the prototype Gorgon Stare system, begin to generate ever-larger amounts of usable information. This reconnaissance package incorporates five electro-optical and four infrared cameras that send images to a ground control station (GCS) that combines them into a single composite view, expanding the search zone and enhancing the images at the same time (Hagerman 2009). The autonomous operation of the system has the potential to not only reduce the number of crew members required to operate the Global Hawk but also to avoid some of the pitfalls that other UAVs have demonstrated. In particular, when control links have been lost to Predators and Reapers, the aircraft have responded with unpredictable behavior. Some have returned to their points of origin, or to the location where the data link was lost, but others have simply continued to fly in the direction they were headed when the signal was lost, continuing blithely along until lack of fuel caused a crash or until impact with a terrain feature such as a mountaintop.

One drawback of the Global Hawk is its high cost. A key attraction of many UAV systems is their relative affordability, but a fully outfitted Global Hawk costs more than $35 million.

While nowhere near the cost of a new F-22 Raptor, it is still a figure that gives pause to appropriations officials. Although lower-cost, lower-capability systems have been considered expendable, or at least attritable, the same cannot be said of Global Hawk. Despite its ability to fly above many conventional air defenses, Global Hawk is vulnerable to the systems fielded by first-rate military powers. For that reason, it is outfitted with an array of electronic detection gear, jammers, a laser warning system, and a long-range radar receiver, as well as a towed decoy that serves as a last-ditch defense against a missile attack. The entire defensive package serves to increase its survivability if it is employed against a peer competitor nation, but it is likely that Global Hawks would not provide substantial assistance in such a conflict.

The primary operator of the Global Hawk is the U.S. Air Force, which plans to continue purchasing new units until at least 2015. Global Hawk will completely replace the U-2 spy plane in 2015, when the older aircraft will formally retire (Majumdar 2011). The U.S. Navy also flies Global Hawks, including a special variant, the MQ-4C, designed for broad area maritime surveillance. NATO has agreed to purchase eight Global Hawks for coalition use, and NATO member Germany has purchased five more for its Luftwaffe. Other nations have expressed interest in possibly purchasing Global Hawks, including Australia, Canada, India, Japan, New Zealand, and South Korea. The U.S. National Air and Space Administration (NASA) has obtained three of the original Global Hawk prototypes for scientific purposes, including environmental testing and monitoring weather patterns (Daly 2009, 342).

MARCbot

The MARCbot (Multi-function Agile Remote Control Robot) is one of the simplest yet most effective robots currently in use by American forces in Iraq. The system was designed by Exponent Engineering and Scientific Consulting (EESC) in direct response to requests by soldiers deployed to Iraq (Clifton 2005). The robot is a small wheeled vehicle with a single extendable arm holding a camera. It can be used to peer inside vehicles or buildings that are suspected of containing hazardous materials. The MARCbot is a truly expendable platform, as it costs less than $10,000 per unit. It is easily controlled by a single operator and can run for six hours on a single set of rechargeable batteries. The batteries are a standard military model, reducing supply and recharging problems. The entire robot is only 19 inches wide and 13 inches high, and weighs only 25 pounds. The camera can be extended up to 24 inches above the chassis and is a simple low-light camera (Exponent 2010).

The original proof-of-concept version of the MARCbot represented a simple case of off-the-shelf engineering that used components already readily available but combining them in a new fashion. It included parts of remote-control cars, including the batteries (which lasted only 20 minutes before needing a recharge) and the suspension (mounted on four relatively large wheels). The chassis of the toy was replaced by a wooden platform, which supported an arm module connected only by Velcro. It was almost ludicrous in its simplicity, and it obviously would not withstand the operational environment found in Iraq and Afghanistan. However, it did prove that an improvised explosive device (IED)–spotting robot does not need to be large, heavy, or full of specialized electronics. The MARCbot serves the useful function of being the first line of inspection of suspicious devices—it can be sent to examine a package, and then a bomb-disposal team can be dispatched, if necessary (*Defense Industry Daily*, 2006).

American personnel have a long history of modifying their equipment to suit their needs in the field, and the MARCbot exemplifies this habit. Not only was the system designed using input from soldiers in the field, some enterprising soldiers have turned it into an offensive weapon. One unit created its own version of a mobile IED by attaching a Claymore antipersonnel mine to the front of a MARCbot, ready to be detonated if the MARCbot operator spotted any insurgents in the unit's route. While not the ideal use of the robot, considering the likely destruction of the delivery mechanism in addition to the target, this modification did follow the pattern established by EECS in designing and refining the system to fit the operational needs of troops in the field (Singer 2009, 32).

MULE

The Multifunctional Utility/Logistics and Equipment vehicle (MULE) is a six-wheeled ground vehicle designed by and produced by Lockheed Martin for the U.S. Army. It is designed to support infantry forces by carrying equipment, disabling mines, and in some cases, mounting weapons as large as an antitank gun (Singer 2009, 113). It senses the terrain using a LADAR (laser radar) that creates a three-dimensional view of the robot's surroundings and then maps a route to follow based on the most efficient way to avoid obstacles in its path. It also uses thermal scanners to help distinguish between humans, vehicles, and other obstacles (Govers 2008).

Each of the MULE's wheels has a separate in-hub motor and an articulated suspension. This allows each wheel to move at a different speed and to traverse extremely rough terrain, such as that likely to be found in an urban combat environment. The system was specifically designed to support the Army's brigade combat team (BCT) organizational model. It is large enough, at 2.5 tons, to carry a substantial amount of cargo for troops, yet small enough to be sling-loaded under UH-60 Blackhawk helicopters or internally transported by C-130 and C-47 cargo airplanes. These transport options allow the MULE to operate in support of air assault units, which have typically been limited to light weaponry as a means of remaining mobile, leaving them inherently vulnerable to heavily armed or armored forces.

In December 2009, the U.S. Army cancelled variants of the system designed for specialization in transport and countermine operations. The decision signaled the Army's plan to focus primarily on the weapons-carrier version of the MULE, the XM 1219 Armed Robotic Vehicle. While this system can still be pressed into a cargo transportation role, its integrated sensor package and weapon mounts make it less functional in that capacity than the originally planned XM 1217 Transport (Lockheed Martin 2006).

PackBot

"PackBot" is a generic term applied to an entire series of military robots produced by the iRobot Corporation. It is a lightweight, easily controlled ground vehicle designed for a wide variety of military and civilian missions. More than 2,500 PackBots have been used by U.S. military units in Iraq and Afghanistan. The most prominent military role that PackBots have played thus far has been in the field of detecting and disabling improvised explosive devices (IEDs), but they have also been used at vehicle and personnel checkpoints, in passive surveillance locations, and to locate enemy snipers through acoustic location packages (iRobot 2011, 2).

PackBots are relatively light, weighing between 40 and 60 pounds. They can be guided by wire or radar and can be outfitted with video cameras to allow the operator to remain in a protected position. They have a low center of gravity and sit on tracks capable of climbing stairs. Thus far, the robots have proven extremely durable, although they are certainly not indestructible. Their relatively low cost makes them somewhat expendable, particularly if the alternative is to send a human explosive ordnance expert into a dangerous situation (iRobot 2011, 4).

PackBot has a number of standard variations, each built on the same basic platform but specialized for a number of essential missions. This reflects iRobot founder Rodney Brooks's vision of robotic design: rather than a single system with a massive number of capabilities but also an enormous price tag, he advocates a much larger quantity of smaller robots with reduced abilities but specialized functions. This allows redundancy of capabilities without the vulnerability of losing everything in a single system, but it can also prove frustrating for operators in complex environments, where more than one specialization might be required during a single operation.

The PackBot Scout is a ground ISR platform of the most basic type. It has several payload bays that can be equipped with sensors. It is rugged enough to be dropped six feet onto a concrete surface without taking damage, which has led troops to utilize the PackBot Scout as a means to look around corners or inside buildings. The PackBot can be hurled through an open window or over a low wall and commanded to scan the interior of buildings without exposing human team members to hostile fire.

The PackBot Explorer is a more dedicated ISR platform, and it is also more specialized than the Scout. While the main unit can absorb substantial punishment, the same is not always true of the sensory payload. The PackBot Explorer has an extendable camera head with laser pointers, microphones, and other detection equipment. It can navigate extremely difficult terrain with relative ease, allowing it to peek into potentially hostile areas in advance of troop movements. It can also be left in a stable position to passively observe its surroundings, a function that greatly preserves the battery life of the unit.

PackBot EOD has fewer sensors but includes a remote-manipulation arm to allow an operator to move objects and potentially disarm or disable explosive devices (Singer 2009, 111). It has proven extremely effective in reducing the IED threat in Iraq, particularly in the urban sectors of Baghdad once considered the most dangerous regions of the country. It has also gained substantial public recognition through its relatively negative depiction in Kathryn Bigelow's 2008 film *The Hurt Locker.* In the film, the robots appear unreliable and the subjects of the film, explosive ordnance disposal (EOD) technicians in Baghdad, prefer to disarm bombs by hand rather than waiting for a remote option. In reality, EOD teams in Iraq and Afghanistan view approaching bombs as the last resort, and bomb-makers have gone to great lengths to construct bombs that robots cannot tamper with.

The most recent base model of the PackBot is the PackBot 510. This model uses a hand controller similar to home video game consoles, on the theory that this system will prove intuitive for younger operators. The PackBot 510 is designed for modularity, allowing reconfiguration of the robots through the exchange of various kits. These include an EOD kit, a first responder kit, a hazardous materials detection module, a separate explosives sniffer system, a sniper detection kit that uses multiple microphones to triangulate the origin of a gunshot, and a fast tactical maneuvering kit designed to streamline IED detection by allowing the PackBot to keep pace with dismounted infantry forces (iRobot 2011, 5–6).

The effectiveness of PackBots has been demonstrated almost constantly on the modern battlefield. Perhaps the greatest testament to their utility came from the enemy. On a number of occasions, insurgents have specifically targeted the robots that had proven so successful in foiling intended attacks. Brookings Institution researcher P. W. Singer investigated the relationships the develop between operators and their robots, and reported that American soldiers have tended to anthropomorphize their PackBots, attributing personality quirks and behaviors to their machines. Some units have named their PackBots and essentially incorporated them into the unit, and the loss of PackBots, particularly in combat, has led to soldiers exhibiting the signs of grief over the loss of their "friends." (Singer 2009, 337–338). The PackBots have become so ubiquitous that iRobot Corporation has provided an in-theater "robot hospital" to repair damaged units and return them more quickly to service. Several units beyond repair have received military "funerals" from their human comrades. In 2011, iRobot volunteered PackBots for service exploring the Fukushima Daiichi nuclear plant in Japan to check radiation levels and conditions, relieving humans of the dangerous task of entering a nuclear reactor devastated by an earthquake and tsunami in 2011.

Pioneer

The RQ-2 Pioneer is a small battlefield UAV that was initially designed for artillery spotting but eventually evolved into a variety of roles. It was created by a joint venture between Israel's IAI (Israeli Aerospace Industries) and the American company AAI (formerly Aviation Armaments, Inc.). The system was first adopted in 1986 and remained a vital UAV platform for the U.S. Army, Marine Corps, and Navy for more than two decades before finally being retired in 2007.

The Pioneer has a relatively compact 17-foot wingspan, a 14-foot length, and a 40-inch height. It can use both electro-optical and infrared sensors, and recent upgrades allow the transfer of images from the aircraft to a ground control station via real-time downlink. It can be remotely flown via line-of-sight or satellite data link, but it is also capable of semiautonomous flight following pre-programmed coordinates. The Pioneer can be launched in a variety of ways, including a conventional runway takeoff, a pneumatic catapult launch, or a rocket-assisted takeoff. It can land on a runway, it has an arrester hook assembly for rapid deceleration, or it can be flown into a shipboard net for recovery. While airborne, the Pioneer has a range of 115 miles, a service ceiling of 15,000 feet, and a maximum payload of 100 pounds (U.S. Navy 2009).

The Pioneer represents a rare case of the United States purchasing and adopting an advanced military system from a foreign developer. It was initially envisioned by IAI, building on the success of the earlier Scout and Mastiff systems. Eventually, IAI formed a joint company, (Pioneer UAV, in partnership with AAI) to more efficiently market the system to the U.S. military, particularly the Navy. The Navy had expressed reservations about purchasing UAVs from foreign suppliers but thought the concept of a forward aerial observer to direct naval gunfire was ideal to keep battleships, with their massive guns, relevant in the modern era of aircraft carriers and cruise missiles. Although Pioneer UAV is based in Maryland, the home of AAI, the vast majority of the UAV's components, including the vital sensor pods, are produced in Israel.

Pioneers were first placed on American battleships in 1986, where they assisted in spotting targets for the ships' main batteries. However, like most successful UAV systems, their operators soon found new roles for them to fulfill, and they were almost immediately pressed into mine-seeking duties over the Persian Gulf. The Marine Corps used its first Pioneers to designate targets for manned, fixed-wing aircraft, while the Army first used Pioneers to check the safety of flight routes used by Apache helicopters. Soon, Pioneers had been fitted with meteorological instruments, chemical detection gear, and communication interception gear, rapidly demonstrating the versatility of UAVs and whetting the services' appetites for more unmanned platforms, particularly ones specially designed for difficult or dangerous missions.

The Pioneer proved to be an extremely durable, reliable system. It performed reconnaissance missions in support of U.S. operations in Iraq during Operation Desert Storm. During that conflict, a Pioneer flying over Iraqi lines became the first UAV in history to receive an enemy surrender attempt, as images of Iraqi troops waving white flags were transmitted to American battleships engaged in bombardment (U.S. Navy 2009; Singer 2009, 57). While the Pioneer had no means of accepting the surrender, its operators were able to halt the attack and send a shore party to accept the surrender. Pioneers were used extensively during the ill-fated humanitarian mission to Somalia in 1993, where they accomplished a sortie rate of nearly 93 percent and demonstrated that they were dependable platforms. Pioneers accompanied troop deployments to Bosnia and Kosovo later in the decade but had less luck in these theaters when faced by an enemy with extensive air defenses. Of the five Pioneers performing ISR over Kosovo during the 79-day bombardment in 1999, only one survived the conflict. The remainder were shot down or lost due to weather or operator errors. When American forces led a coalition invasion of Iraq in 2003 during Operation Iraqi Freedom, Pioneers once again assisted in scouting the layout of Iraqi air defenses and the dispositions of Iraqi troops. Marine units flying Pioneers flew nearly 400 sorties in the initial weeks of the invasion, amassing 1,300 combat hours and losing only one aircraft. Pioneers continued to fly over Iraq and Afghanistan until 2007, averaging 7,000 combat hours per year before being retired in favor of the RQ-7 Shadow, also produced by AAI.

Predator

The General Atomics RQ-1/MQ-1 Predator is probably the best known and most successful unmanned aerial vehicle to date. It is currently flown by the U.S. Air Force; the Central Intelligence Agency (CIA); the militaries of Great Britain, Italy, and Turkey, and other U.S. government agencies associated with homeland security, border patrol, and law enforcement. It has flown in military missions in the Balkans, Afghanistan, Iraq, Pakistan, Yemen, and Libya, first as an ISR platform (the RQ-1) and later as an armed attack UAV (the MQ-1). The successful use of Predator systems has greatly accelerated the demand for unmanned combat aerial vehicles (UCAVs), leading directly to the development of the RQ-9/MQ-9 Reaper and driving a host of research and development programs in this field.

The Predator is characterized by an aerodynamic design that allows for extreme fuel efficiency. The wingspan of over 50 feet, length of 27 feet, and height of less than seven feet create a sleek, low-slung profile and make the aircraft seem much smaller on the ground. In reality, its dimensions are comparable to the P-51 Mustang of World War II fame, although the Predator weighs only a fraction of the famous escort fighter's mass. Although the aircraft has a maximum speed of 135 miles per hour, its typical cruising speed is closer to 90 miles per hour. A single turboprop engine, mounted at the rear of the aircraft, essentially pushes the Predator through the air and is so efficient that a single tank of fuel can last 24 hours, providing a range of 2,300 miles. The quiet engine is inaudible from two miles away, allowing the craft to fly at its service ceiling of 25,000 feet with minimal chance of detection by humans on the ground (Air Combat Command 2010; Daly 2009, 295–299).

Predator's greatest value comes from the wide variety of sensor packages it can wield. They range from simple video cameras, including the one used by the pilot to fly the aircraft, to infrared cameras, synthetic aperture radars, and laser target designators. In many ways, a Predator provides many of the same capabilities as the most sophisticated surveillance satellites but can do so at a much lower cost and on demand, allowing tremendous battlefield awareness for commanders.

Predators can be controlled using either a line-of-sight system or an encrypted satellite feed. The takeoff and recovery functions are typically performed by a team of on-site operators, referred to as a Landing and Recovery Element (LRE). Once the aircraft is airborne, control is transferred to the Mission Control Element (MCE), who will pilot the aircraft and operate its devices from a ground control station (GCS). The GCS is small enough to be transported in one piece on C-130 Hercules cargo plane, and the Predator aircraft can be disassembled into six pieces for rapid deployment and reassembly. In theory, all of the elements needed for Predator operations can be transported, assembled, and in action in less than 24 hours. Predator crews include a pilot to fly the aircraft and a sensor operator to perform the ISR and targeting aspects of the mission. Any Predator attack requires a positive action by the pilot; the system is not configured for autonomous hostile action. However, the plane is capable of autonomous flight, particularly in the event that communication with the GCS is lost, and it can be preprogrammed to fly via waypoints and to return to base if the signal from the GCS disappears. The Predator can navigate using inertial navigation or a GPS system.

Flying Predators has long been likened by the media to playing a video game, but the U.S. Air Force has long maintained a policy that certified pilots make the most successful Predator pilots. Many of the same skills apply, and the best operators are reported to be those who can envision themselves actually inside the aircraft. This is particularly important in emergency situations, when the pilot's training may allow avoidance of accidents that would occur if the operator had no experience at the controls of a manned aircraft. While the control system is not identical to a manned system, many of the individual components are similar, also contradicting the video game analogy. Some operators in training have reported frustration with the system due to a time lag between making a flight correction and the subsequent response from the UAV. Also, the onboard pilot camera provides a much narrower field of vision than a cockpit, causing one pilot to refer to the experience of flying the Predator as "like trying to fly while looking through a soda straw" (Martin and Sasser 2010, 24). Nevertheless, Predators and other UAVs have proven so successful that the U.S. Air Force is currently training more unmanned aircraft pilots than manned aircraft pilots each year.

Although the U.S. military has used drones in a variety of roles for nearly 100 years, including as aerial targets, reconnaissance platforms, and guided bombs, the Predator represented a revolutionary step forward in drone technology, ironically by eschewing many of the elements commonly associated with advanced airframe technology. Previous reconnaissance drones, such as the Teledyne Ryan Fire Fly and Lightning Bug, relied on speed to penetrate enemy airspace, moving rapidly over well-defended terrain before enemy units could organize a response. While this allowed some locations to be scouted on missions too dangerous for piloted aircraft, the speed of such drones hindered the resolution of their imagery, and their fuel consumption was so high that most missions were of limited range and endurance. Also, the loud engines and large radar signature made covert surveillance with such drones impractical, if not impossible.

In the 1980s, the CIA became interested in the possibility of smaller, low-profile aerial surveillance systems. In part, this reflected an important lesson of the Vietnam War: while the U.S. Air Force preferred fast attack jets for ground attack missions, such aircraft were ineffective for forward air controllers (FACs). Instead, FACs preferred slower propeller-driven aircraft that could loiter over an area for a substantial period. The slower speed allowed a much better vantage point for artillery spotting, bomb damage assessments, and surveillance gathering. Such aircraft were also easier to fly and cheaper to maintain. However, while rugged, they proved susceptible to ground fire, as antiaircraft batteries could easily track and fire on the slower aircraft. As such, they were most effective in a permissive air environment and presented the possibility of unacceptably high casualty rates if flown against a peer competitor such as the Soviet Union.

By creating an unmanned propeller-driven airframe, many of the risks associated with the craft disappeared, as did some of the costs. If such a system could be built cheaply enough to be effectively expendable, it might provide a perfect ISR platform, even in a nonpermissive environment. Plus, without the need to lift a human pilot, the aircraft could be scaled down, reducing both costs and radar cross-sections. The addition of nonmetallic composite materials for construction, a new aerodynamic airframe for maximum endurance, and a specially designed quiet engine to reduce the aural indicators completed the physical design of the Predator. When coupled with new sensors and a remote control system, the latest ISR drone made its maiden flight in 1994.

After an initial demonstration of the new system, General Atomics Aeronautical Systems, Inc. (GA-ASI), received a contract to develop and produce the Predator. By mid-1995, the system had proven so effective in testing that it was designated for deployment to Gjadër Airbase in Albania, where the craft overflew the Balkan region to monitor the conflicts arising from the dissolution of the Yugoslavian state. Two of the Predators were lost during these operations, and others experienced mechanical and operator difficulties. Nevertheless, they proved invaluable in documenting human rights atrocities committed during the brutal civil war.

In 1999, Predators again deployed to the Balkan region, this time in support of the aerial attacks against Serbia. In this conflict, Predators provided ISR, used laser targeting to assist in the delivery of precision munitions dropped by manned aircraft, and performed bomb damage assessments. The Serbian army possessed a respectable air defense system but managed to shoot down only two Predators, both in May. A third aircraft went down due to extreme icing conditions, prompting a redesign of the system to allow the addition of automatic deicing equipment to the leading edges of the Predators' wings. Several more Predators experienced mechanical and operator failures, further illustrating that the technology required further refining before it could be committed against a first-class opponent. However, the Predators once again proved their worth, and General Atomics received further orders to expand the number of UAVs in both the CIA and U.S. Air Force inventories.

In 2000, the CIA conducted a short series of Predator reconnaissance flights over Afghanistan. On at least two occasions, a Predator sent images of a man suspected to be Al Qaeda's leader, Osama bin Laden. In early 2001, the U.S. Air Force conducted live fire tests to see if Predators could be armed with AGM-114C Hellfire missiles. These weapons, previously utilized primarily by U.S. Army attack helicopters, could be outfitted with a variety of warheads, including antipersonnel high explosive variants designed to rain deadly shrapnel across a broad area, or armor-piercing warheads initially designed to penetrate Soviet armored vehicles but also suitable for attacking protected buildings. The tests demonstrated the efficacy of arming the drones, allowing an immediate response if a high-value target was spotted during an ISR mission. A human operator would still need to give the aircraft permission to fire, as at no point did the Predator receive autonomous firing capability.

In the aftermath of the September 11, 2001, attacks, Predators flew some of the earliest missions over Afghanistan, beginning on September 18, 2001. Armed Predators began operations on October 7 and have since launched hundreds of Hellfire missiles against Al Qaeda and Taliban targets. In addition to armed reconnaissance and laser designation, Predators have also been utilized in a close air support role to assist American and allied units in close combat. Predator missions have also flown over Pakistan, including missile attacks against Al Qaeda forces. Predators provided some surveillance of sites suspected of harboring bin Laden, including the Abottabad compound where he was located and killed on May 2, 2011. Similar ISR and attack flights against Al Qaeda–affiliated groups in Yemen have also been reported since 2002.

In the period from 1991 to 2003, American and coalition military forces maintained two "no-fly zones" over northern and southern Iraq. Unmanned aircraft played an increasingly important role in the overwatch missions. Several Predators were engaged by Iraqi air defenses, and at least two were shot down. More served to entice Iraqi aircraft to violate the zones, operating essentially as aerial targets. In late 2002, a Predator fitted with AIM-92 Stinger missiles attacked an Iraqi MiG-25. In the ensuing engagement, the manned fighter shot down the much slower and lighter-armed Predator, the first air-to-air shootdown of a UAV in combat.

When coalition forces launched a ground invasion of Iraq in 2003, Predators played a vital scouting role. Although conventional operations ended quickly, the subsequent insurgency led to the utilization of Predators in overwatch positions, detecting and often engaging insurgents, particularly those involved in planting improvised explosive devices or launching mortar attacks against coalition units. Predators soon became the most-requested aerial asset in Iraq and Afghanistan. Predators had flown more than 250,000 hours by June 2007 and doubled that time by February 2009. By mid-2010, Predators accounted for over a million hours of flight (General Atomics 2010; U.S. Air Force 2009).

Predators provided ISR and attack capabilities in Libya since the beginning of American and NATO aerial operations (Shanker 2011). Their usage was controversial, as was the entire involvement of NATO in an internal Libyan conflict. They have also been used in antipiracy roles, as their extremely sensitive detection gear, long loiter time, and immunity to the weapons possessed by Somali pirates make them ideal for patrolling the waters off of the African continent (Axe 2009). Although the U.S. Air Force accepted delivery of its final Predator system in March 2011, it is likely that other nations and American agencies will continue to field the system for the foreseeable future (General Atomics 2011).

Raven

The AV (formerly AeroVironment) RQ-11 Raven is a small, tactical UAV designed for use by ground forces needing an ISR platform to survey their immediate environment. It is one of the most-produced UAVs in history, with approximately 10,000 copies in use by American forces and thousands more units delivered to allied military units. It is a simple, dependable, and effective platform that has proven extremely popular with troops in the field, particularly those in urban environments (Piper 2005). The Raven has been repeatedly upgraded, utilizing suggestions made by military personnel who have used the unit in the field, thus allowing the system to be much more in line with their actual needs rather than the perceptions of corporate engineers working far from the combat environment.

The Raven has a 55-inch wingspan and a length of 36 inches, but its weight, at just over four pounds, is sufficiently light that a user can launch via a throw. It has a small electric motor that can reach a maximum speed of 60 miles per hour, yet it is virtually silent when in operation. This powerful motor also makes the Raven capable of operation in high crosswinds, a constant problem for most light observation craft. It cruises at 35 miles per hour and has an endurance of up to 90 minutes. Its range is just over six miles, more than sufficient for most tactical applications but limiting its effectiveness as an artillery spotter or forward observer. It typically flies only a few hundred feet above ground level, but unlike most tactical UAVs, the Raven has a relatively high service ceiling and can fly at altitudes of up to 10,000 feet, which is particularly important when it is being utilized in mountainous regions such as Afghanistan, where American forces often engage the enemy at extreme altitudes (Air Force Special Operations Command 2010; Daly 2009, 261–263).

Despite its small size, the Raven has a relatively sophisticated guidance system. While it can be remotely operated by an individual using a ground control station that is essentially a modified laptop, the Raven is also capable of completely autonomous flight. It can be preprogrammed to follow GPS waypoints, allowing operators to focus on the images being transmitted back by its electro-optical or infrared sensors. Reflecting the short-term nature of its usage, and the need for ground teams to remain in motion, the Raven has a single-button recall function that orders it to return to its point of origin. Of course, a remote operator can also designate a different landing point for the aircraft. To land, the Raven descends into a deep stall, which can be considered akin to a controlled crash. However, the airframe is remarkably rugged and reusable for up to 100 missions before the wear and tear of usage will render it inoperable.

The Raven requires almost no maintenance beyond recharging its power source and selecting which sensor package should be used. It can be easily disassembled for rapid transport and deployment, with a single soldier able to carry all of the equipment necessary for a Raven sortie. At $35,000 per aircraft, each Raven costs less than 1 percent of the larger and better-known Predator. This low price reflects a desire by the U.S. Army, the primary user of these drones, to avoid the "gold plating" that has encumbered so many other UAVs, adding sensory capabilities far beyond the needs of troops in the field and correspondingly increasing the price of each unit until it ceases to be an expendable system. For the time being, the Raven is truly expendable, as each unit costs approximately the same amount as a single Joint Direct Attack Munition (JDAM) dropped from an Air Force or Navy aircraft (Boeing 2011, 2). Of course, troops in the field take pains to maintain their supply of drones, but they do not risk human lives trying to recover downed aircraft.

The Raven is popular not only with American forces; it is one of the most exported UAV systems in history. Currently, at least 11 other nations (Australia, the Czech Republic, Denmark, Estonia, Iraq, Italy, Lebanon, the Netherlands, Norway, Spain, and the United Kingdom) possess and operate Ravens within their military forces. The Raven's system is so basic and straightforward that the U.S. Department of Defense sees no threat in the proliferation of this system, although some upgrades to the platform are currently reserved for American users. The Raven is widely considered one of the most useful and successful UAVs currently fielded by U.S. military forces in Afghanistan and Iraq, and will likely remain a mainstay of ground units in the foreseeable future.

Reaper

The General Atomics RQ-9/MQ-9 Reaper largely resembles a much larger version of the company's Predator remotely piloted vehicle. It has a much more powerful engine, wider wingspan, greater speed and service ceiling, and higher payload capacity. While still fundamentally an ISR platform, the Reaper has much greater utility than the Predator as an attack aircraft.

After initial tests of armed Predators proved successful, General Atomics Aeronautical Systems, Inc. (GA-ASI) decided to design and develop a much larger variant, even before the military requested such a device. Because the company's leadership believed that such a request would be almost inevitable but would also be accompanied by an invitation to other firms to compete for a production contract, GA-ASI decided that having a demonstration model ready much earlier than any other competitors would provide a distinct advantage in the future bidding process. The decision had substantial inherent risks, not least of which was that the military would begin to turn away from unmanned aircraft, but in this case, it paid off handsomely for the company.

In 2001, GA-ASI began flight testing the new system, which it called the Predator B. Design engineers differed on what configuration would work best for the propulsion system, and the company produced prototypes using both turboprop and jet engines. The jet version had a greater speed and ceiling but a much shorter endurance and lighter payload, causing the U.S. Air Force to select the turboprop model for further development. To some observers, this represented a fundamental shift in the typical behavior of Air Force acquisitions officers, who had previously been enamored with the notion of a jet-powered UAV.

The Reaper has a 66-foot wingspan, only slightly larger than that of the Predator. However, its 950-horsepower engine provides a cruising speed of over 240 miles per hour, nearly three times faster than its predecessor. It can fly more than twice as high, with a ceiling slightly above 50,000 feet, and it can fly for 35 hours without refueling when not heavily laden with munitions. The armed version can carry 800 pounds of weaponry in an internal bay and a further 3,000 pounds on external hardpoints. This means the Reaper can be armed with up to 14 AGM-114 Hellfire missiles, or it can carry laser-guided bombs or Joint Direct Attack Munitions (JDAMS) using its central hardpoints (Air Combat Command 2010a; Daly 2009, 295–299).

The Reaper has proven extremely successful in combat usage, prompting the Air Force to contract for more than 300 aircraft in addition to the necessary ground control stations, supporting software, and maintenance supplies. It was first deployed to Iraq and Afghanistan in 2007, and quickly took over many of the roles already being performed by the Predator systems (*Air Force Times*, 2007). Because the Reaper is substantially faster, it is able to respond to requests from ground units much quicker. Its larger supply of munitions make it better at performing hunter-killer missions for longer periods, without having to land and rearm on a continual basis.

The command systems for Reapers are similar to Predators, with most American remote pilots operating out of Creech Air Force Base in Nevada. Like Predators, Reapers have been adapted for maritime patrols over the coastal United States and near eastern Africa. They are also utilized in border control missions along U.S. boundaries with Canada and Mexico. Reapers have been exported to the military forces of Australia, Italy, Turkey, and the United Kingdom.

Given the success of the Reaper, it was almost inevitable that General Atomics would continue to develop its unmanned vehicles to capitalize on the changing needs of the U.S. military. One of the chief complaints about the Reaper is that its speed, while greater than the Predator's, is still limited by the use of a single turboprop engine. Another is that the system could be made significantly more stealthy, reducing its vulnerability. To address these concerns, GA-ASI designed the Avenger, a turbofan-powered, armed UAV akin to the Reaper but much faster and stealthier. The new aircraft, which made its maiden flight in the spring of 2009, can carry the same amount of weaponry as the MQ-9 but in an internal weapons bay, which reduces the aircraft's radar signature. Its exhaust heat is sent through a series of baffles, making it far less obvious to infrared sensors such as those found in the warheads of many air-to-air and surface-to-air missiles. Further enhancing its capabilities, the Avenger has a Lynx synthetic aperture radar and an automatic targeting and ranging system. A naval version, the Sea Avenger, can take off from and land on an aircraft carrier, and has foldable wings to facilitate storage in the onboard hangar. This system, first flown in 2010, has reignited naval interest in GA-ASI systems.

ScanEagle

The ScanEagle is a UAV that combines low costs, long endurance, and the ability to engage in limited swarm behavior into a single package. It began as the product of a collaboration between Insitu Group and Boeing, with Boeing eventually purchasing Insitu but keeping it largely independent. The ScanEagle is an outgrowth of the SeaScan program, which was begun in 2000. This system, designed for maritime tactical surveillance, sought to minimize the area needed for launch, control, and recovery of a small UAV. The SeaScan proved so versatile that vessels as small as 20 feet in length can use it. The first flight of the ScanEagle occurred in 2002, allowing its designers to benefit from lessons learned from previous UAV designs. It requires less logistical support than most UAVs of a similar size, as it can run on gasoline rather than a unique or difficult-to-acquire fuel. Its capabilities are similar to the RQ-1 Predator, in that it cruises at 70 miles per hour and can remain aloft for more than 30 hours. However, it has a much smaller visual and radar signature, as its wingspan is only 10 feet, while the Predator's is over five times wider. Further, it can be launched from a small SuperWedge system, making its employment far easier in difficult conditions far from airstrips. An air-launched variant of the ScanEagle is currently in development. Its recovery also does not require an airstrip; rather, it utilizes a new SkyHook method.

The ScanEagle has hooked wingtips, allowing it to snag a rope suspended from a 15-meter pole. These launch and recovery methods have made the ScanEagle attractive to the U.S. Navy, its primary customer, as it can be used by virtually any surface vessel, including the new Littoral Combat Ship (LCS) (Insitu 2010).

The ScanEagle's modular sensor system allows the operator to choose from a variety of sensory packages. The conventional selections include a digital camera with a 25x zoom or an infrared camera capable of 7.5x magnification. As of 2008, ScanEagle can be equipped with an extremely small synthetic aperture radar (SAR), providing greater utility in poor weather or over a smoke-covered area (Boeing 2008). It has also been tested with laser designators, biochemical detection gear, and magnetometers. In 2007, Boeing began testing the ScanEagle with a ShotSpotter sniper detection package, enhancing the usefulness of the system in the protection of fixed installations or in an oversight capacity above troops in the field (Boeing 2007).

The ScanEagle was initially designed to fly with limited autonomy, in that it could follow preprogrammed routes using GPS waypoints, or it could be "locked on" to a moving target that it would autonomously follow. In 2005, the ScanEagle received a software upgrade that enabled it to generate its entire flight path independent of external guidance. This capability became increasingly important as users wished to control multiple aircraft from a single ground control station (GCS). The current configuration of the GCS has two consoles, each of which can simultaneously control four aircraft. In July of 2011, Boeing successfully tested the ability of UAVs to use information from other UAVs to map and fly a search pattern. Further complicating the operation, Boeing used two ScanEagles and one Procerus Unicorn, supplied by the Johns Hopkins University Applied Physics Laboratory. The three aircraft created waypoints, flight paths, and terrain maps that they shared, allowing a rapid search of a rugged, mountainous region. They simultaneously sent their combined data to a GCS.

The U.S. Navy and Marine Corps have used ScanEagles extensively in Iraq and the Persian Gulf. They have also been utilized in antipiracy operations off the coast of Somalia, where they have maintained overwatch above the key shipping lanes of the region and tracked suspected pirate ships back to their coastal hideouts (Axe 2009). In July 2011, Insitu announced that ScanEagles had flown over 500,000 hours of combat missions. Systems have also been purchased by the U.S. Air Force for use in base security operations. ScanEagles have been exported to Australia, Canada, Colombia, the Netherlands, Poland, and the United Kingdom.

Insitu has taken the experience gained from the ScanEagle project to develop two more generations of UAVs. The Insight has a dual payload bay system, allowing it to carry two sensor systems at once, but otherwise it closely resembles the ScanEagle in size and performance, including using the same SuperWedge and SkyHook launch and recovery systems. The Integrator, first demonstrated in 2007, is slightly larger than its predecessors but is modularized for easy transport and assembly by a single operator. It can carry all of the same payloads but also has a fuselage bay that can carry intelligence collection, communication, or weapons packages. It is also equipped with external hardpoints, which might be used for Spike missiles, making it a true miniaturized competitor to the Predator.

Shadow

The RQ-7 Shadow is a tactical UAV developed by the AAI Corporation (formerly Aircraft Armaments, Inc.) for use primarily by the U.S. Army and Marine Corps. It is designed to replace the RQ-2 Pioneer and conforms to the Army's requirements that the aircraft have short takeoff and landing requirements, a gasoline-powered engine, sufficient range to serve as an artillery target designator, and at least a four-hour loiter time. The Army also desired a platform that could carry both electro-optical and infrared sensors in a single device, rather than requiring a choice between the two prior to any mission.

With a wingspan of 14 feet, length of 11 feet, and a height of 40 inches, the dimensions of the Shadow are similar to the Pioneer. Likewise, the service ceiling (15,000 feet), range (68 miles), cruising speed (100 miles per hour), and endurance (six hours) are almost identical to the earlier UAV. The Shadow also uses the same sensor package as the latest Pioneer upgrades (AAI Corporation 2009). The similarities have caused some critics to wonder why a new system was necessary, or even desirable. The Shadow is lighter than the Pioneer and can be launched from a trailer using a hydraulic catapult. It has a stronger engine than the Pioneer but ironically carries a lighter payload. Its power plant generates four times more watts, but current configurations do not require such amounts. In many ways, the Shadow represents more of a software upgrade than a hardware improvement. The Shadow does have a significant advantage in its data link capabilities. In an effort to standardize equipment as much as possible, the Army has incorporated a tactical common data link system for control of UAVs. The Shadow is capable of autonomous flight using GPS navigation to preprogrammed checkpoints.

Like the Pioneer, the Shadow has proven attractive to international purchasers. In addition to the U.S. Army, Marine Corps, and Navy, the Shadow is currently flown by the military forces of Australia, Italy, Pakistan, Poland, Romania, South Korea, Sweden, and Turkey. It has been flown in support of operations in Afghanistan and Iraq, logging more than 100,000 combat hours of flight over Iraqi airspace alone. Because the Pioneer was formally retired from U.S. military service in 2007, the Shadow stands in position to be the Army's workhorse battlefield surveillance system for the near future and may compete to fill a similar role if NATO chooses to standardize its UAV equipment (Daly 2009, 248–250).

TALON

TALONs are versatile tracked ground robots developed by Foster-Miller and now produced by QinetiQ. More than 3,000 TALON systems have been sent into combat zones, primarily Iraq and Afghanistan. TALONs are a durable unit capable of traversing almost any terrain. According to QinetiQ, they have no difficulty in sand or snow, or under up to 100 feet of water, and they can climb stairs and move across damaged urban areas littered with rubble (QinetiQ 2011).

TALONs are extremely portable, as even the heaviest variants weigh only slightly more than 100 pounds. They are controlled by radio signals or fiberoptic cables from a single operator's computer unit. The TALON is equipped with multiple sensors, including a color video camera with infrared and night vision modes. Its endurance is particularly high for its class of robots, as it can function at normal speeds for more than eight hours on a single charge and can remain in standby mode for a week on a single battery charge. Foster-Miller, a company dedicated to the engineering aspects of its product rather than the pure scientific innovation, prides itself on the simplicity and durability of its systems (Krasner 2007).

The reconnaissance version of TALON is extremely fast, in part because the chassis is a lighter version weighing only 60 pounds. It can wield a variety of sensors to perform its mission depends entirely on the human operator for decision making. The EOD version of TALON has a manipulator arm to aid the operator in accessing and disabling explosive devices. While TALON is not an expendable machine, at less than $250,000 apiece, it is still a far less expensive option than having humans manually disable bombs. Further, the rugged design has allowed many EOD TALONs to withstand the blast of IEDs or to be quickly repaired and returned to service (Krasner 2007). EOD TALONs have proven so effective that jihadists have begun to actively target the systems, hoping destruction of the robots will allow attacks against human EOD technicians. To counter this problem, Foster-Miller established a repair bay in Baghdad, designed to quickly return robots to duty (Singer 2009, 110–113).

The HAZMAT TALON carries sensors designed to test for chemical, temperature, and radiation conditions, and display those results in real time for the operator. Predecessors of the current HAZMAT system contributed to recovery operations at the World Trade Center in the aftermath of the September 11, 2001, attacks and proved capable of long operations in contaminated zones. Many withstood repeated decontaminations without degrading their function. Because the robots are able to traverse such rough terrain, they were invaluable in providing advanced warning of dangerous locations within the remnants of the fallen towers (QinetiQ 2011).

Perhaps the most radical of the TALON robots is the version long envisioned by literature and science fiction films. The SWORDS (Special Weapons Observation Reconnaissance Detection System) variant is a TALON that can be equipped with a variety of small arms, ranging from an M-16 carbine to a four-shot 66-millimeter incendiary rocket, the M202A1 Flash. While

the SWORDS appears fearsome, the TALON is not currently configured for autonomous operation. As such, the system requires a
dedicated operator and cannot fire without human permission.
However, it has proven disturbingly accurate due to a number
of factors. The TALON SWORDS has a much greater visual acuity
than a human shooter, even one armed with enhanced optics. The
digital camera is capable of extreme magnification and is directly
integrated with the weapon barrel, meaning the shots will go precisely where the camera is aimed. Because the firing platform is
completely stable, the most common causes of missed shots,
including breathing, muscle twitches, and blinking at the wrong
time, are for the most part eliminated. The machine is immune
to the nervous effect of being fired on by the enemy, and in fact
can use incoming fire to determine the position of hostile targets.
Because the weapons are computer controlled, even a TALON
SWORDS with a mounted machine gun can be tasked to fire a single round, turning even the most rapid-fire weapons into virtual
sniper platforms. When a .50 caliber M82 Barrett rifle is mounted,
the SWORDS is capable of hitting a human-sized target more than
two miles away with a single shot (Jewell 2004; McElroy 2007).

After more than 10 years of military deployments, the
TALON systems have earned rave reviews from their users,
prompting Foster-Miller to expand production to meet demand.
As a subsidiary of QinetiQ, Foster-Miller has access to more resources and more clients, all interested in the remarkable success
of the TALON line. The company has begun to develop TALON's
successor, the Modular Advanced Armed Robotic System
(MAARS), a larger and more adaptable system with increased
autonomous capability.

Warrior

Following the success of iRobot's PackBot, the company chose to
create a larger system capable of carrying substantially more
cargo into a combat environment. The resulting robot, Warrior,
maintains many of the same design elements of the PackBot in a
more robust, faster system and has been described by one
observer as "essentially a PackBot on steroids" (Singer 2009, 24).
It has a pair of dual tracks for propulsion, allowing it to climb
steps and slopes with ease. Not only is the robot body stronger,

with a payload capacity of 200 pounds, but the manipulator arm is much enhanced and can carry objects weighing 150 pounds. This allows the Warrior to carry a PackBot into a hostile situation and lift it into position, possibly by placing it through a window (Brandon 2009).

While most variants of Warrior are nonlethal, including explosive ordnance disposal, ISR, and casualty extraction models, iRobot has partnered with Metal Storm to weaponize the system, and the sheer size of Warrior makes it likely to be able to mount fairly substantial weapons systems for infantry support. It has a modular payload system, and includes standard power and Ethernet interfaces on the chassis. The Warrior has a built-in obstacle detection system, making navigation easier for the operator, who drives the robot using a handheld controller and laptop computer (iRobot 2010).

Documents
Missile Technology Control Regime

The Missile Technology Control Regime (MCTR) is a voluntary international association designed to limit the number and type of delivery mechanisms for weapons of mass destruction by prohibiting transfers of technology that would enable states to develop advanced weaponry. Although designed primarily to limit cruise missile technology, many of the provisions of the MTCR could potentially be applied to other forms of remotely piloted or autonomous aircraft, including many of the most advanced UAVs in operation today, as the only exceptions within the MTCR guidelines are for manned aircraft. Begun by the leading members of NATO in 1987, the MTCR has now expanded to 34 partner nations, including several former members of the Warsaw Pact.

Guidelines for Sensitive Missile-Relevant Transfers, 1987

1. The purpose of these Guidelines is to limit the risks of proliferation of weapons of mass destruction (i.e., nuclear, chemical and biological weapons), by controlling transfers that could make a contribution to delivery systems (other than manned aircraft) for such weapons. The Guidelines are also intended to limit the risk of controlled items and their technology falling into the hands of terrorist groups and individuals. The Guidelines are

not designed to impede national space programs or international cooperation in such programs as long as such programs could not contribute to delivery systems for weapons of mass destruction. These Guidelines, including the attached Annex, form the basis for controlling transfers to any destination beyond the Government's jurisdiction or control of all delivery systems (other than manned aircraft) capable of delivering weapons of mass destruction, and of equipment and technology relevant to missiles whose performance in terms of payload and range exceeds stated parameters. Restraint will be exercised in the consideration of all transfers of items within the Annex and all such transfers will be considered on a case-by-case basis. The Government will implement the Guidelines in accordance with national legislation.

2. The Annex consists of two categories of items, which term includes equipment and technology. Category I items, all of which are in Annex items 1 and 2, are those items of greatest sensitivity. If a Category I item is included in a system, that system will also be considered as Category I, except when the incorporated item cannot be separated, removed or duplicated. Particular restraint will be exercised in the consideration of Category I transfers regardless of their purpose, and there will be a strong presumption to deny such transfers. Particular restraint will also be exercised in the consideration of transfers of any items in the Annex, or of any missiles (whether or not in the Annex), if the Government judges, on the basis of all available, persuasive information, evaluated according to factors including those in paragraph 3, that they are intended to be used for the delivery of weapons of mass destruction, and there will be a strong presumption to deny such transfers. Until further notice, the transfer of Category I production facilities will not be authorised. The transfer of other Category I items will be authorised only on rare occasions and where the Government (A) obtains binding government-to-government undertakings embodying the assurances from the recipient government called for in paragraph 5 of these Guidelines and (B) assumes responsibility for taking all steps necessary to ensure that the item is put only to its stated end-use. It is understood that the decision to transfer remains the sole and sovereign judgement of the Government.

3. In the evaluation of transfer applications for Annex items, the following factors will be taken into account:

   A. Concerns about the proliferation of weapons of mass destruction;

B. The capabilities and objectives of the missile and space programs of the recipient state;

C. The significance of the transfer in terms of the potential development of delivery systems (other than manned aircraft) for weapons of mass destruction;

D. The assessment of the end use of the transfers, including the relevant assurances of the recipient states referred to in sub paragraphs 5.A and 5.B below;

E. The applicability of relevant multilateral agreements.

F. The risk of controlled items falling into the hands of terrorist groups and individuals.

4. The transfer of design and production technology directly associated with any items in the Annex will be subject to as great a degree of scrutiny and control as will the equipment itself, to the extent permitted by national legislation.

5. Where the transfer could contribute to a delivery system for weapons of mass destruction, the Government will authorize transfers of items in the Annex only on receipt of appropriate assurances from the government of the recipient state that:

A. The items will be used only for the purpose stated and that such use will not be modified nor the items modified or replicated without the prior consent of the Government;

B. Neither the items nor replicas nor derivatives thereof will be re transferred without the consent of the Government.

6. In furtherance of the effective operation of the Guidelines, the Government will, as necessary and appropriate, exchange relevant information with other governments applying the same Guidelines.

7. The Government will:

A. provide that its national export controls require an authorisation for the transfer of non-listed items if the exporter has been informed by the competent authorities of the Government that the items may be intended, in their entirety or part, for use in connection with delivery systems for weapons of mass destruction other than manned aircraft;

B. and, if the exporter is aware that non-listed items are intended to contribute to such activities, in their entirety or part, provide, to the extent compatible with national export controls, for notification by the exporter to the authorities referred to above, which will decide whether or not it is

> appropriate to make the export concerned subject to
> authorisation.

8. The adherence of all States to these Guidelines in the interest of
    international peace and security would be welcome.

Source: Military Technology Control Regime. *Guidelines for Sensitive
Missile-Relevant Transfers.* Retrieved from http://www.mtcr.info/
english/guidetext.htm.

National Defense Authorization, Fiscal Year 2001

*National Defense Authorization bills are relatively routine, albeit enor-
mous, legislative pieces that establish broad directives and funding limits for the U.S. Armed Forces. Often, they are used to place extremely
specific requirements upon the services and to tie those requirements to
financial resources. In 2001, the authorization included orders that the
Armed Forces should gradually increase the size and capabilities of their
unmanned systems. In part, this was due to resistance from all four services to the idea of unmanned platforms. The goals established by this section of the act were ambitious, but by some definitions the first goal has
been met and efforts to reach the second remain underway. By forcing
each service to report its research activities, Congress opened the possibility of joint programs to eliminate duplication of efforts and to share
the benefits of the individual research initiatives, while reaffirming the
central role of DARPA in the production of unmanned vehicles.*

Sec. 220. Unmanned Advanced Capability Combat Aircraft and Ground Combat Vehicles

(a) Goal.—It shall be a goal of the Armed Forces to achieve the field-
     ing of unmanned, remotely controlled technology such that—

   (1) by 2010, one-third of the aircraft in the operational deep
        strike force aircraft fleet are unmanned; and

   (2) by 2015, one-third of the operational ground combat
        vehicles are unmanned.

(b) Report on unmanned advanced capability combat aircraft and
     ground combat vehicles.—(1) Not later than January 31, 2001,
     the Secretary of Defense shall submit to the congressional
     defense committees a report on the programs to demonstrate
     unmanned advanced capability combat aircraft and ground

combat vehicles undertaken jointly between the Director of the
Defense Advanced Research Projects Agency and any of the
following:

(A) The Secretary of the Army.

(B) The Secretary of the Navy.

(C) The Secretary of the Air Force.

(c) Funds.—Of the amount authorized to be appropriated for
Defense-wide activities under section 201(4) for the Defense
Advanced Research Projects Agency, $100,000,000 shall
be available only to carry out the programs referred to in
subsection (b)(1).

Source: National Defense Authorization, Fiscal Year 2001. Public Law
106-398, Oct 30, 2000. Retrieved from http://www.dod.mil/dodgc/olc/
docs/2001NDAA.pdf.

International Code of Conduct against Ballistic Missile Proliferation, 2002

The International Code of Conduct against Ballistic Missile Prolifera-
tion is an international agreement that seeks to prevent the acquisition
of intermediate-range ballistic missiles by nations that do not currently
possess such technology. Currently, 130 nations have agreed to the pro-
visions of the code. The treaty is important related to UAVs because
many of the most high-technology systems might technically qualify as
ballistic missiles capable of delivering weapons of mass destruction, even
if they are remotely piloted aircraft.

Preamble

The Subscribing States:

Reaffirming their commitment to the United Nations Charter;

Stressing the role and responsibility of the United Nations in the
field of international peace and security;

Recalling the widespread concern about the proliferation of
weapons of mass destruction and their means of delivery;

Recognizing the increasing regional and global security challenges
caused, inter alia, by the ongoing proliferation of Ballistic Missile
systems capable of delivering weapons of mass destruction;

Seeking to promote the security of all states by fostering mutual trust through the implementation of political and diplomatic measures;

Having taken into account regional and national security considerations;

Believing that an International Code of Conduct against Ballistic Missile Proliferation will contribute to the process of strengthening existing national and international security arrangements and disarmament and non-proliferation objectives and mechanisms;

Recognising that subscribing States may wish to consider engaging in co-operative measures among themselves to this end;

1. Adopt this International Code of Conduct against Ballistic Missile Proliferation (hereinafter referred to as "the Code");
2. Resolve to respect the following Principles:
    1. Recognition of the need comprehensively to prevent and curb the proliferation of Ballistic Missile systems capable of delivering weapons of mass destruction and the need to continue pursuing appropriate international endeavours, including through the Code;
    2. Recognition of the importance of strengthening, and gaining wider adherence to, multilateral disarmament and non-proliferation mechanisms;
    3. Recognition that adherence to, and full compliance with, international arms control, disarmament and non-proliferation norms help build confidence as to the peaceful intentions of states;
    4. Recognition that participation in this Code is voluntary and open to all States;
    5. Confirmation of their commitment to the United Nations Declaration on International Cooperation in the Exploration and Use of Outer Space for the Benefit and in the Interest of All States taking into particular Account the Needs of Developing Countries, adopted by the United Nations General Assembly (Resolution 51/122 of 13 December 1996);
    6. Recognition that states should not be excluded from utilising the benefits of space for peaceful purposes, but that, in reaping such benefits and in conducting related cooperation, they must not contribute to the proliferation of Ballistic Missiles capable of delivering weapons of mass destruction;

7. Recognition that Space Launch Vehicle programmes should not be used to conceal Ballistic Missile programmes;

8. Recognition of the necessity of appropriate transparency measures on Ballistic Missile programmes and Space Launch Vehicle programmes in order to increase confidence and to promote non-proliferation of Ballistic Missiles and Ballistic Missile technology;

3. Resolve to implement the following General Measures:

   1. To ratify, accede to or otherwise abide by:

      • the Treaty on Principles Governing the Activities of States in the Exploration and Use of Outer Space, including the Moon and Other Celestial Bodies (1967),

      • the Convention on International Liability for Damage Caused by Space Objects (1972), and

      • the Convention on Registration of Objects Launched into Outer Space (1975);

   2. To curb and prevent the proliferation of Ballistic Missiles capable of delivering weapons of mass destruction, both at a global and regional level, through multilateral, bilateral and national endeavours;

   3. To exercise maximum possible restraint in the development, testing and deployment of Ballistic Missiles capable of delivering weapons of mass destruction, including, where possible, to reduce national holdings of such missiles, in the interest of global and regional peace and security;

   4. To exercise the necessary vigilance in the consideration of assistance to Space Launch Vehicle programmes in any other country so as to prevent contributing to delivery systems for weapons of mass destruction, considering that such programmes may be used to conceal Ballistic Missile programmes;

   5. Not to contribute to, support or assist any Ballistic Missile programme in countries which might be developing or acquiring weapons of mass destruction in contravention of norms established by, and of those countries' obligations under, international disarmament and non-proliferation treaties;

4. Resolve to implement the following:

   1. Transparency measures as follows, with an appropriate and sufficient degree of detail to increase confidence and to promote non-proliferation of Ballistic Missiles capable of delivering weapons of mass destruction:

1. With respect to Ballistic Missile programmes to:
   - make an annual declaration providing an outline of their Ballistic Missile policies.Examples of openness in such declarations might be relevant information on Ballistic Missile systems and land (test-) launch sites;
   - provide annual information on the number and generic class of Ballistic Missiles launched during the preceding year, as declared in conformity with the pre-launch notification mechanism referred to hereunder, in tiret iii);

2. With respect to expendable Space Launch Vehicle programmes, and consistent with commercial and economic confidentiality principles, to:
   - make an annual declaration providing an outline of their Space Launch Vehicle policies and land (test-) launch sites;
   - provide annual information on the number and generic class of Space Launch Vehicles launched during the preceding year, as declared in conformity with the pre-launch notification mechanism referred to hereunder, in tiret iii);
   - consider, on a voluntary basis (including on the degree of access permitted), inviting international observers to their land (test-) launch sites;

3. With respect to their Ballistic Missile and Space Launch Vehicle programmes to:
   - exchange pre-launch notifications on their Ballistic Missile and Space Launch Vehicle launches and test flights. These notifications should include such information as the generic class of the Ballistic Missile or Space Launch Vehicle, the planned launch notification window, the launch area and the planned direction;

Subscribing States could, as appropriate and on a voluntary basis, develop bilateral or regional transparency measures, in addition to those above.

Implementation of the above Confidence Building Measures does not serve as justification for the programmes to which these Confidence Building Measures apply;

Organisational aspects

Subscribing States determine to:

Hold regular meetings, annually or as otherwise agreed by Subscribing States;

Take all decisions, both substantive and procedural, by a consensus of the Subscribing States present;

Use these meetings to define, review and further develop the workings of the Code, including in such ways as:

- establishing procedures regarding the exchange of notifications and other information in the framework of the Code;

- establishing an appropriate mechanism for the voluntary resolution of questions arising from national declarations, and/or questions pertaining to Ballistic Missile and/or Space Launch Vehicle programmes;

- naming of a Subscribing State to serve as an immediate central contact for collecting and disseminating Confidence Building Measures submissions, receiving and announcing the subscription of additional States, and other tasks as agreed by Subscribing States; and

- others as may be agreed by the Subscribing States, including possible amendments to the Code.

Source: Hague Code of Conduct website: http://www.hcoc.at/background docs.php

Department of Defense Reports to Congress on Development and Utilization of Robotics and Unmanned Ground Vehicles, October 2006

On a regular basis, Congress requires the Department of Defense to supply reports regarding specific weapons systems, plans, and performance on the battlefield. At times, these reports assess the technological progress of the United States in comparison to that of peer competitors within a specific area of interest. In this report, the Office of the Under Secretary of Defense for Acquisition, Technology, and Logistics prepared a report called for in the National Defense Authorization Act for Fiscal Year 2006, showing how robotics had been incorporated into American military operations thus far. This section of the report compares American research and development with that of many of its closest allies.

(10) An assessment of international research, technology, and military capabilities in robotics and unmanned ground vehicle systems.

A number of U.S. allies currently conduct research and development (R&D) activities directed towards developing military capabilities for robotics and unmanned ground vehicles (UGVs). Canada conducts research in the areas of autonomous systems with a focus on sensors and integration for robotic systems, control systems for robotic applications, data communication systems, robotic vehicle platforms, artificial intelligence for robotic systems, and the ergonomic aspects of man-machine interface. Germany has sponsored Science and Technology (S&T) efforts directed towards the development of critical technologies for UGVs including perception, intelligent control, autonomous robotic vehicle platforms, as well as human interface and planning. Recently, it has begun to focus on the development of small (i.e., man packable) robots. Australia is concentrating on the areas of platform-related technologies systems and weapons, man-unmanned systems, control theory and control systems.

France is focusing on the areas of system collaboration, weapons, level of autonomy, and night vision and electronic sensors to include countermine and de-mining technologies. The United Kingdom is primarily working on navigation, mobility, communication, and ground vehicle integration. Israel is conducting work on tank systems dealing with laser range-finders, design and fabrication of tank systems. South Korea recently initiated research focused on development of a Multi-function Utility/Logistics Equipment (MULE)-like platform, as well as real time tracking and Human Robotic Interface (HRI) efforts which they hope will ultimately result in a vehicle that can be used to monitor the Demilitarization Zone. Other international efforts include HRI by Switzerland and systems for mine clearing and mobility by Denmark. In summary, the aforementioned countries are concentrating on capabilities for urban operation and combat application, as opposed to Japan, where defense applications for robotic technologies are their primary goal.

In general, U.S. capabilities, research, and technologies are leading the way for the international efforts. However, Japan's effort with HRI is comparable, while the humanoid-like robotic technology may be somewhat ahead of those in the U.S. at present. South Korea began investing heavily in HRI and may partner with the U.S. in the future. Canada is increasing their investing efforts with platforms and may be considered comparable to U.S. platform technology.

Source: Office of the Under Secretary of Defense, Acquisition, Technology, and Logistics, Portfolio Systems Acquisition, Land Warfare, and Munitions, Joint Robotics Enterprise. *Report to Congress:*

Development and Utilization of Robotics and Unmanned Ground Vehicles (October 2006). Retrieved from http://www.ndia.org/Divisions/ Divisions/Robotics/Documents/Content/ContentGroups/Divisions1/ Robotics/JGRE_UGV_FY06_Congressional_Report.pdf

U.S. Department of State Legal Advisor Harold Koh, Speech to the American Society of International Law Annual Conference, 2010

The legality of drone strikes has been a hotly debated topic during the Global War on Terror. In particular, the question of whether individuals can legally be targeted, even if they are not engaged in active military operations. The case of Anwar al-Awlaki, in particular, aroused special interest in the United States because he was an American citizen serving as a religious leader within Al Qaeda. Awlaki's father filed a lawsuit seeking to remove his son's name from the targeted killing list, but the lawsuit was dismissed. In September of 2011, a pair of Predators fired Hellfire missiles at Awlaki's convoy in Yemen, killing the radical cleric. Harold Koh's remarks to the American Society of International Law offer the Obama administration's rationale behind using remotely piloted aircraft for targeted killings. Koh argues that the decision to use UAV strikes to target leaders of Al Qaeda does not violate constitutional protections or the laws of armed conflict, although he makes no mention of the special conditions that arise when a U.S. citizen is the object of the attack.

"... [I]t is the considered view of this administration ... that targeting practices, including lethal operations conducted with the use of unmanned aerial vehicles (UAVs), comply with all applicable law, including the laws of war. ... As recent events have shown, Al Qaeda has not abandoned its intent to attack the United States, and indeed continues to attack us. Thus, in this ongoing armed conflict, the United States has the authority under international law, and the responsibility to its citizens, to use force, including lethal force, to defend itself, including by targeting persons such as high-level al Qaeda leaders who are planning attacks. ... [T]his administration has carefully reviewed the rules governing targeting operations to ensure that these operations are conducted consistently with law of war principles, including:

- First, the principle of *distinction*, which requires that attacks be limited to military objectives and that civilians or civilian objects shall not be the object of the attack; and

- Second, the principle of *proportionality*, which prohibits attacks that may be expected to cause incidental loss of civilian life, injury to civilians, damage to civilian objects, or a combination thereof, that would be excessive in relation to the concrete and direct military advantage anticipated.

In U.S. operations against al Qaeda and its associated forces—including lethal operations conducted with the use of unmanned aerial vehicles—great care is taken to adhere to these principles in both planning and execution, to ensure that only legitimate objectives are targeted and that collateral damage is kept to a minimum.

[S]ome have suggested that *the very use of targeting* a particular leader of an enemy force in an armed conflict must violate the laws of war. But individuals who are part of such an armed group are belligerent and, therefore, lawful targets under international law.... [S]ome have challenged *the very use of advanced weapons systems*, such as unmanned aerial vehicles, for lethal operations. But the rules that govern targeting do not turn on the type of weapons system involved, and there is no prohibition under the laws of war on the use of technologically advanced weapons systems in armed conflict—such as pilotless aircraft or so-called smart bombs—so long as they are employed in conformity with applicable laws of war.... [S]ome have argued that the use of lethal force against specific individuals fails to provide adequate process and thus constitutes *unlawful extrajudicial killing*. But a state that is engaged in armed conflict or in legitimate self-defense is not required to provide targets with legal process before the state may use lethal force. Our procedures and practices for identifying lawful targets are extremely robust, and advanced technologies have helped to make our targeting even more precise. In my experience, the principles of distinction and proportionality that the United States applies are not just recited at meeting. They are implemented rigorously throughout the planning and execution of lethal operations to ensure that such operations are conducted in accordance with all applicable law.... Fourth and finally, some have argued that our targeting practices violate *domestic law*, in particular, the long-standing *domestic ban on assassinations*. But under domestic law, the use of lawful weapons systems—consistent with the applicable laws of war—for precision targeting of specific high-level belligerent leaders when acting in self-defense or during an armed conflict is not unlawful, and hence does not constitute 'assassination.'"

Source: American Society of International Law. 2010. *US State Dept Legal Adviser Lays Out Obama Administration Position on Engagement, "Law of 9/11."* Retrieved from http://www.asil.org/files/KohatAnMtg100325.pdf.

Aircraft Procurement Plan Fiscal Years (FY) 2012–2041, U.S. Air Force, 2011

Modern weapons procurement cycles can last decades, and with the enormous costs associated with fielding new systems, it is necessary to plan as far into the future as possible. Periodically, Congress requires the Department of Defense to produce procurement plans for years or decades in advance. In this plan, the Department of Defense (DoD) demonstrated its understanding of the need to develop and procure unmanned systems to absorb some of the workload from piloted platforms. Of particular interest is the plan to create a "centerpiece" long-range strike bomber capable of operating as both a manned and unmanned platform, despite the obvious drawbacks to such a dual-use aircraft.

II. Summary of the Aircraft Investment Plan

In keeping with the Department's desire to provide a flexible and balanced force, the aviation plan provides the diverse mix of aircraft needed to carry out the six joint missions identified above. The very nature of modern warfare makes categorizing aircraft into bins of like capability more and more difficult. When considering aviation investment plans, the Department must increasingly consider the potential complimentary capabilities resident in the cyber and space domains, as well as across other aircraft types.

The capabilities provided by aircraft identified in this plan translate into four principal investment objectives that were reflected in the FY 2011 budget and are carried forward in the FY 2012 budget request and the FY 2012–2041 aviation plan:

- **Meet the demand for persistent, unmanned, multirole intelligence, surveillance, and reconnaissance (ISR) capabilities.** The number of platforms in this category—R/MQ-4 Global Hawk-class, MQ-9 Reaper, and MQ-1 Predator-class unmanned aircraft systems (UAS)—will grow from approximately 340 in FY 2012 to approximately 650 in FY 2021. This 90 percent capacity increase will be effectively multiplied by capability improvements afforded by the acquisition of vastly improved sensors and the replacement of Air Force MQ-1s with more capable MQ-9s. This capacity increase of Air Force MQ-1B and MQ-9 platforms will enable the establishment of 65 orbits by the end of FY 2013. In addition to funding the MQ-4C Broad Area Maritime Surveillance (BAMS) aircraft, the Navy is in the early stages of developing an Unmanned Carrier Launch Airborne Surveillance and Strike (UCLASS) system to provide persistent ISR and precision strike. Though omitted from the procurement and inventory statistics in this report, the Army will buy 78 MQ-1C Gray Eagle UASs between FY 2012 and FY 2016. Procurement plans of a subset of these aircraft—MQ-9, USMC Group 4 UAS, UCLASS, and a possible follow-on UAS—are less specific after FY 2016 to allow flexibility to continue growth as required.

- **Modernize long-range strike (LRS) capabilities.** The enduring need for long-range attack capabilities will be met by a combination of current and future aircraft and weapons systems. The current fleet of Air Force bombers continues to be modernized so that it can retain long range strike capabilities through the 2030s. To deter and defeat anti-access threats, DoD is creating an LRS family of systems with a new penetrating, nuclear capable bomber program as the centerpiece. The new bomber will be designed to accommodate manned or unmanned operations.

Source: *Air Force Magazine*. website: http://www.airforce-magazine .com/SiteCollectionDocuments/Reports/2011/May%202011/Day25/ AircraftProctPlan2012-2041_052511.pdf

Policy Options for Unmanned Aircraft Systems, Congressional Budget Office, 2011

The nonpartisan Congressional Budget Office (CBO) exists to provide independent analysis of budget proposals and their likely economic effects for the U.S. Congress. The CBO does not make policy recommendations; rather, it simply handles the analysis of different outcomes based on budgetary decisions. In 2011, the CBO prepared a report showing how UAVs would affect defense budgets, based on the various proposals laid out by each service. The summary of the budgetary analysis includes a synopsis of the current UAV acquisitions of each of the services, as well as a brief explanation of the services' desires for future UAV purchases.

The Air Force's Plans

The Air Force currently operates at least four medium-sized or large unmanned aircraft: Global Hawks, Predators, Reapers, and Sentinels. The largest aircraft is the jet-powered RQ-4 Global Hawk, and the Air Force has 14 of them, according to [the] CBO's information. The most numerous, at approximately 175 aircraft, is the MQ-1 Predator, a piston-engine propeller aircraft that can take still or video imagery and shoot Hellfire missiles. A larger version of the Predator, the turboprop-powered MQ-9 Reaper, is beginning to enter the force, and about 40 have been delivered as of 2011 . . . The RQ-170 Sentinel is a stealthy reconnaissance aircraft whose existence has only recently been acknowledged by the Air Force. Most performance characteristics of the Sentinel remain classified.

The Air Force's near-term goals are to increase the number of Global Hawk and Reaper aircraft that can be continuously and simultaneously operated. To meet that goal, the Air Force plans to purchase 288 Reapers (48 per year from 2011 through 2016) and 28 Global Hawks from 2011 through 2018 . . . The Air Force is also exploring the characteristics that would be desired in a larger aircraft a generation beyond the Global Hawk.

About $20.4 billion will be needed for the aircraft the Air Force plans to purchase through 2020, CBO estimates: $7.3 billion for Global Hawks and $13.1 billion for Reapers and their follow-on. Costs would average about $2.0 billion per year through 2020.

The Army's Plans

The Army currently operates three medium-sized unmanned aircraft systems: Hunters, Shadows, and Predators. Overall, the Army's inventory includes about 20 MQ-5B Hunters (older aircraft scheduled for retirement by 2013), about 450 RQ-7 Shadows, and about 40 MQ-1 Predators in two versions (specifically, MQ-1 Warrior Alphas and MQ-1C Gray Eagles).

Over the next five years, the Army plans to purchase 20 Shadows to replace losses, upgrade the existing shadows with tactical data links and a laser targeting system, and purchase 107 more of the medium-altitude Grey Eagles. [The] CBO estimates that those plans will cost about $5.9 billion: $1.9 billion for the Shadows and $4.0 billion for the additional Grey Eagles. In the longer term, the Army is exploring concepts for an aircraft that has greater endurance (that is, can stay in the air for a longer time). It also may decide to resume efforts to increase the capabilities of unmanned aircraft used by combat brigades; those plans were shelved when the Army's Future Combat System was cancelled in 2009.

The Navy and Marine Corps' Plans

The Navy is currently testing two new types of aircraft that it hopes to field in the near future—the long-endurance Broad Area Maritime Surveillance (BAMS) aircraft, which is a Global Hawk variant optimized for naval operations, and the MQ-8B Fire Scout unmanned helicopter. The Navy plans to purchase 36 BAMS aircraft at a cost of about $9.4 billion by 2020 and operate them from a few bases worldwide to provide surveillance of activities on the oceans. The Navy also plans to purchase 61 Firescouts by 2020 at a cost of $1.0 billion; those helicopters will be based on selected surface ships and will provide local reconnaissance and the capability to attack hostile surface targets. The Navy's plans call for purchasing a total of 65 BAMS through 2026 and 168 Firescouts through 2028.

The Marine Corps is in the process of fielding the Shadow to support ongoing operations in Southwest Asia. Thirteen systems (with four aircraft per system) had been delivered by the end of calendar year 2009. The Marine Corps does not plan to purchase additional Shadow systems but instead will spend about $120 million to upgrade some Shadows already in its inventory.

In the longer term, the Navy is exploring concepts for a carrier-based unmanned aircraft, called the Unmanned, Carrier-Launched Airborne Surveillance and Strike aircraft, and is currently flying a demonstrator aircraft to help develop the technologies and procedures needed to operate such an aircraft. The Marine Corps is exploring concepts for medium-sized system (currently referred to as the Group 4 Unmanned Aircraft System) that would be designed to perform various missions in support of amphibious operations. Both systems might enter service by 2020.

Source: Congressional Budget Office Pub No 4083. Retrieved from http://www.cbo.gov/sites/default/files/cbofiles/ftpdocs/121xx/doc12163/06-08-uas.pdf.

References

AAI Corporation. 2009. *Shadow 600*. Retrieved from http://www.aaicorp.com/pdfs/shadow600_12-18-09bfinal.pdf.

AeroVironment. 2011. *Dragon Eye Datasheet*. Retrieved from http://www.avinc.com/downloads/Dragon_Eye_AV_datasheet.pdf.

Air Combat Command, Public Affairs Office. 2009. *RQ-4 Global Hawk*. Retrieved from http://www.af.mil/information/factsheets/factsheet.asp?id=13225.

Air Combat Command, Public Affairs Office. 2010. *MQ-1B Predator*. Retrieved from http://www.af.mil/information/factsheets/factsheet.asp?fsID=12.

Air Combat Command, Public Affairs Office. 2010a. *MQ-9 Reaper*. Retrieved from http://www.af.mil/information/factsheets/factsheet.asp?fsID=6405.

Air Force Special Operations Command, Public Affairs Office. 2010. *RQ-11B Raven*. Retrieved from http://www.af.mil/information/factsheets/factsheet.asp?fsID=10446.

Air Force Times. 2007. "Reaper Scores Insurgent Kill in Afghanistan" (October 29). Retrieved from http://www.airforcetimes.com/news/2007/10/airforce_mq9_reaper_071029w/.

Axe, David. 2009. "4 Fronts for Pirate-Navy Battles as Pirate Attacks Continue." *Popular Mechanics*. (October 1). Retrieved from http://www.popularmechanics.com/technology/military/4285201.

Boeing. 2007. *Boeing to Test Sniper Fire Detection and Location Technology for U.S. Air Force* (January 23, 2007). Retrieved from http://www.boeing.com/news/releases/2007/q1/070123a_nr.html.

Boeing. 2008. *Boeing Flight-Tests 2-Pound Imaging Radar Aboard ScanEagle Unmanned Aircraft*. (March 18). Retrieved from http://www.boeing.com/news/releases/2008/q1/080318a_nr.html.

Boeing Defense, Space & Security. 2011. *Joint Direct Attack Munition (JDAM) Backgrounder*. http://www.boeing.com/defense-space/missiles/jdam/docs/jdam_overview.pdf.

Brandon, Alan. 2009. "iRobot Warrior 700 designed to deliver . . . More Robots." *Gizmag* (August 26). Retrieved from http://www.gizmag.com/irobot-warrior-700/12624/.

Clifton, Matthew. 2005. "Tiny Robot Carries Big Responsibility in Iraq." *Transformation* (September 14). Retrieved from http://www.defense.gov/transformation/articles/2005-09/ta091405a.html.

Crane, David. 2005. "USMC Dragon Runner Mini Recon Robot (Ground Robot)/UGV for Urban Warfare Ops." *Defense Review* (January 14). Retrieved from http://www.defensereview.com/us-marine-corps-experimenting-with-new-recon-robot-for-urban-warfare-ops/.

Daly, Mark, ed. 2009. *Jane's Unmanned Aerial Vehicles and Targets*. Alexandria, VA: Jane's Information Group.

Defense Industry Daily. 2006. "Mighty Mites: MARCbots Add Exponent to IED Land-Mine Detection" (June 1). Retrieved from http://www.defenseindustrydaily.com/mighty-mites-marcbots-add-exponent-to-ied-landmine-detection-updated-02314/.

Exponent. 2010. *MARCbot*. Retrieved from http://www.exponent.com/marcbot_product/.

General Atomics. 2010. *Predator-Series UAS Family Achieves One Million Flight Hours* (April 6). Retrieved from http://www.ga.com/news.php?read=1&id=284&page=2.

General Atomics. 2011. *Air Force Accepts Delivery of Last Predator* (March 7). Retrieved from http://www.ga.com/news.php?read=1&id=341.

Govers, Francis X. 2008. "The MULE: Multipurpose Logistics Vehicle." *Robot Magazine* (September/October). Retrieved from http://www.botmag.com/articles/mule.shtml.

Hagerman, Eric. 2009. "Coming Soon: An Unblinking 'Gorgon Stare' For Air Force Drones." *Popular Science Online* (August 26). Retrieved from http://www.popsci.com/military-aviation-amp-space/article/2009-08/coming-soon-unblinking-gorgon-stare-air-force-drones.

Insitu. 2010. *ScanEagle Unmanned Aircraft System*. Retrieved from http://www.insitu.com/documents/Insitu%20Website/Marketing%20Collateral/ScanEagle%20Folder%20Insert.pdf.

iRobot Corporation. 2010. *iRobot 710 Warrior*. Retrieved from http://www.irobot.com/gi/filelibrary/pdfs/robots/iRobot_710_Warrior.pdf.

iRobot Corporation. 2011. *One Robot, Unlimited Possibilities: iRobot 510 PackBot*. Retrieved from http://www.irobot.com/gi/filelibrary/pdfs/robots/iRobot_510_PackBot.pdf.

Jewell, Lorie. 2004. "Armed Robots to March into Battle." *Transformation* (December 6). Retrieved from http://www.defense.gov/transformation/articles/2004-12/ta120604c.html.

Krasner, Jeffrey. 2007. "Robots Going in Harm's Way." *Boston Globe* (March 12). Retrieved from http://www.boston.com/business/technology/articles/2007/03/12/robots_going_in_harms_way/.

Lockheed Martin. 2006. *MULE/ARV-A(L)*. Retrieved from http://www.lockheedmartin.com/data/assets/mfc/PC/MFC_MULE_pc.pdf.

Majumdar, Dave. 2011. "Global Hawk to Replace U-2 Spy Plane in 2015. *Air Force Times* (August 10). Retrieved from http://www.airforcetimes.com/news/2011/08/dn-global-hawk-to-replace-u2-spy-plane-081011/.

Martin, Matt J. and Charles W. Sasser. 2010. *Predator: The Remote-Control Air War over Iraq and Afghanistan: A Pilot's Story.* Minneapolis, MN: Zenith Publishing.

McElroy, Damien. 2007. "Armed Robots Go to War in Iraq." *Telegraph* (London) (August 5). Retrieved from http://www.telegraph.co.uk/news/worldnews/1559521/Armed-robots-to-go-to-war-in-Iraq.html.

Northrop Grumman. 2010. *MQ-8B Fire Scout*. Retrieved from http://www.as.northropgrumman.com/products/mq8bfirescout_navy/assets/firescout-new-brochure.pdf.

Piper, Raymond. 2005. "Small UAV Provides Eyes in the Sky for Battalions." *Army News Service* (February 17). Retrieved from http://www.military.com/NewsContent/0,13319,usa1_021705.00.html.

QinetiQ. 2011. TALON: Fast, Mobile, and Specialized. Retrieved from http://www.qinetiq-na.com/products/unmanned-systems/talon/.

Raibert, Marc, Keven Blankespoor, Gabriel Nelson, Rob Playter, and the BigDog Team. 2008. *BigDog, the Rough-Terrain Quadruped Robot.* Retrieved from http://www.bostondynamics.com/img/BigDog_IFAC_Apr-8-2008.pdf.

Shanker, Thom. 2011. "Obama Sends Armed Drones to Help NATO in Libya War." *New York Times* (April 21). Retrieved from http://www.nytimes.com/2011/04/22/world/africa/22military.html.

Singer, P. W. 2009. *Wired for War: The Robotics Revolution and Conflict in the Twenty-First Century.* New York: Penguin Books.

Stewart, Joshua. 2011. "Navy: UAV Likely Downed by Pro-Gadhafi Forces." *Navy Times* (August 5). Retrieved from http://www.navytimes.com/news/2011/08/navy-fire-scout-likely-shot-down-libya-080511w/.

Tiron, Roxana. 2004. "Marine Gladiator Charges Ahead." *National Defense* (May). Retrieved from http://www.nationaldefensemagazine.org/archive/2004/May/Pages/Marine_Gladiator3573.aspx.

U.S. Air Force. 2009. "Predator Passes 600,000 Flight Hours." *Air Force News* (September 30). Retrieved from http://www.af.mil/news/story.asp?id=123170356.

U.S. Navy. 2009. *RQ-2A Pioneer Unmanned Aerial Vehicle (UAV).* Retrieved from http://www.navy.mil/navydata/fact_display.asp?cid=1100&tid=2100&ct=1.

Wagner, William. 1982. *Lightning Bugs and Other Reconnaissance Drones.* Fallbrook, CA: Aero Publishers.

Wagner, William and William P. Sloan. 1992. *Fireflies and Other UAVs.* Leicester, United Kingdom: Midland Publishing.

7

Directory of Organizations

This chapter provides an overview of some of the most important organizations associated with the design, production, and utilization of military robots and drones. It is broken into three groups: government agencies, research organizations, and defense conglomerates. The government agencies are organizations that are directly funded by the U.S. government, and they often fulfill both a research and small production capacity. The research organizations are a combination of academic research institutions and private companies that engage in both the design and production of robots. The defense conglomerates are some of the largest corporations in the world, most of which rely heavily upon government defense contracts for the majority of their revenues. Although they all maintain research organizations within their corporate hierarchy, they are best known for the actual production of military hardware, and their research efforts are typically in response to specific government bidding opportunities.

Government Agencies

Air Force Research Laboratory

88th Air Base Wing Public Affairs
5215 Thurlow Street, Building 70
Wright-Patterson Air Force Base, OH 45433-5543
Phone: 937-522-3252
Website: www.wpafb.af.mil/AFRL/

The Air Force Research Laboratory was created in 1997 as a means to consolidate several smaller Air Force laboratories into a single location, on an Air Force base near Dayton, Ohio. The AFRL exists to allow the Air Force to conduct independent research into the development of technologies useful for the U.S. military, particularly those related to airpower, space exploration, and cyberspace. Nearly 6,000 employees, 75 percent civilian, comprise the workforce for the AFRL, which has an annual budget of nearly $2 billion. Although the formal organization is relatively recent, it continues the legacy of previous military research organizations that had existed for nearly a century. It provides a vital research and testing center that is independent of the industrial partners who perform a substantial amount of research and development for the military.

The AFRL is broken into several separate subsidiaries that focus upon a single field of airpower development. They include the directorates of air vehicles; directed energy; human performance; information; materials and manufacturing; munitions; propulsion; sensors; and space vehicles. Several of these directorates have headquarters away from Wright-Patterson AFB, to better facilitate their research with less bureaucratic interference from the central organization. It has been associated with virtually every major aerial weapons system in the Department of Defense, including the design and procurement of unmanned aerial vehicles. For the most part, the USAF has remained content to leave the creation of UAV platforms to industry partners, for now, although the AFRL has been heavily involved in the creation and improvement of sensor systems to be deployed on board the vehicles.

Defense Advanced Research Projects Agency (DARPA)

3701 North Fairfax Drive
Arlington, VA 22203-1714
Phone: 703-526-6630
Fax: 703-528-3655
E-mail: darpapublicaffairsoffice@darpa.mil
Website: http://www.darpa.mil

The Advanced Research Projects Agency (ARPA) was founded in 1958 in response to the Soviet Union's launch of *Sputnik*, the first artificial satellite of the earth. The organization was created to prevent strategic surprises arising from technological innovations made by competitors and potential enemies, while at the same time making major research leaps forward to maintain a significant edge over other military forces. It was renamed in 1972 to the Defense Advanced Research Projects Agency (DARPA) to reflect its role in the military procurement system.

DARPA is not part of the traditional military hierarchy, as it reports directly to the office of the secretary of defense. It maintains strong links to leading figures in academia and industry, and rather than maintaining a huge staff of permanent personnel, it creates short-term project teams designed to bring the brightest minds together to tackle a single challenge. As such, DARPA has less than 300 permanent employees, despite commanding a budget of more than $3.2 billion. The organization attempts to anticipate future military needs and then accelerate technological development to meet those needs. Essentially, it aims for revolutionary, rather than evolutionary, change. Once a technology demonstration is complete, DARPA typically passes the concept to other organizations for actual production, development, and refinement.

DARPA has been involved in an enormous number of transformative projects, not all of which had immediate military applications, but many of which have radically altered the manner in which the U.S. military carries out its mission. Perhaps the most well-known DARPA project was the creation of ARPANET, the predecessor to the modern Internet. This was originally envisioned as a means to facilitate the cooperation of DARPA team members who worked at universities and research laboratories across the country. Even though they worked on the same projects, delays often occurred because they could not share their work effectively, until the computer network, using packet switching technology, linked their databases and made transmission of data effective.

DARPA's initial focus was on nuclear and space policy and technology to support each, but its programs gradually broadened to encompass almost any form of revolutionary technological innovation that might have military ramifications. Major computing advances, information processing, and artificial intelligence projects have significantly altered the development of robotic weaponry, and creation of supporting devices, such as state-of-the-art sensor packages, have made autonomous vehicles far more effective. In addition to its own internal projects, DARPA has found creative ways to stimulate research into fields that it considers important for the future. For example, the DARPA Grand Challenge invited competitors to enter a driverless car competition. If an autonomous vehicle could navigate the extremely difficult terrain chosen for the race, the designers would win a $2 million prize. Within two years, the prize had been won, and DARPA estimated that approximately $100 million had been spent by competitors, an enormous payoff for such a small investment.

DARPA is heavily involved in robotics innovation, contributing to dozens of robotic and drone programs. In 2011, DARPA tested the Hypersonic Glider, an unmanned aircraft capable of speeds in excess of Mach 20 that would offer a global strike capability in a matter of hours. While the initial testing was aborted due to sections of the vehicle's skin peeling off, it did manage to attain its top speed in less than three minutes of flight, a clear indicator of a significant leap forward in unmanned vehicle technology. DARPA has also partnered with Boston Dynamics in the creation of BigDog, a robotic prototype designed to carry supplies for dismounted infantry forces. While many of DARPA's projects are extremely classified, the organization continues to produce substantial results in the effort to maintain the American military's technological superiority over any potential foes.

Robotic Systems Joint Project Office

6501 E 11 Mile Road, MS266
Warren, MI 48397
Fax: 586-282-9282
Website: www.rsjpo.army.mil

The Robotic Systems Joint Project Office (RSJPO) is a consolidated Department of Defense attempt to unify robotics research and eliminate duplication of robotic efforts within the military services. It was created in 1988 with the purpose of merging U.S. Army and U.S. Marine Corps research for unmanned ground vehicles, primarily to be used for tactical observation and explosive ordnance disposal. The office has grown to incorporate partnerships with industry and academia to support the development and procurement of robotic ground systems. In addition to building new robotic platforms, RSJPO also provides field support, including in-theater repair facilities, for deployed units using ground robots.

The most prominent robotic systems that have come under RSJPO purview include the well-known Foster-Miller TALON, the iRobot PackBot, and the EECS MARCbot, although there are dozens of other models in current use that were also obtained with RSJPO assistance. The program serves to speed the delivery of robots into the field by projecting future needs, testing advanced concepts, and continually updating feedback from current users in the field, which is then passed back into the development cycle to enhance the next generation of unmanned ground vehicles (UGVs). By the end of April 2012, the RSJPO claimed to have saved nearly 800 lives by using robots to detonate roadside improvised explosive devices (IEDs) that might have otherwise killed explosive ordnance disposal (EOD) technicians.

Research Organizations

Boston Dynamics

78 Fourth Avenue
Waltham, MA 02451-7507
Phone: 781-663-0586
E-mail: info@BostonDynamics.com
Website: http://www.bostondynamics.com

Boston Dynamics is an engineering company that specializes in robotic design, with a particular focus on biomimetics, robots that draw inspiration from biological organisms. The company was founded by Marc Raibert in 1992 in Waltham, Massachusetts. Raibert, who previously taught at both Carnegie Mellon University and the Massachusetts Institute of Technology, is a pioneer in the development of self-balancing robot limbs. Such limbs make legged robots much easier to design, and given that most robots that interact with humans will need to do so in environments that humans have designed for their own comfort, Raibert's focus has made the human-robot partnership of the future far more likely to succeed.

The company is best known for BigDog, a quadruped robot funded by the Defense Advanced Research Projects Agency (DARPA), with assistance from the Foster-Miller Company and the National Aeronautics and Space Administration Jet Propulsion Laboratory. BigDog is essentially designed to serve as a pack robot for infantry squads moving on foot. It has the ability to keep pace with its soldiers, over virtually any terrain they can traverse comfortably, while carrying 350 pounds of supplies. Far less well known, but potentially just as revolutionary, is the RiSE, an insect-based robot that can climb vertical surfaces using micro-claws, in much the same manner as organic insects do. The robot can adjust the shape of its own frame, allowing it to better balance itself using its precarious toeholds. When outfitted with some form of sensor or transmission equipment, RiSE may become the prototype for an entire new line of surveillance robots.

Boston Dynamics has carved out an interesting niche in the testing of new equipment for military personnel through robotic systems. PETMAN is a robot designed to test uniforms, in particular, chemical protection suits. To accurately assess the effectiveness of the suits, the robot is required to completely emulate the movements of a human being in virtually any environment. Further, PETMAN is capable of emitting sweat to test the effects of wearing such a suit in a hazardous environment. Short of actually placing humans in experimental suits and subjecting them to a hazardous environment, PETMAN is the best means currently available to evaluate the efficiency of new gear.

Carnegie Mellon University Robotics Institute

The Robotics Institute
5000 Forbes Avenue
Pittsburgh, PA 15213-3890
Phone: 412-268-3818
Fax: 412-268-6436
E-mail: robotics@ri.cmu.edu
Website: http://www.ri.cmu.edu

The Carnegie Mellon University (CMU) Robotics Institute is one of the largest centers of robotics research and expertise in the world. It is a subdivision of the CMU School of Computer Science. Founded in 1979, the Robotics Institute was the only source in the world for a Ph.D. in robotics for decades. Today, the institute has more than 400 affiliates, a budget of over $50 million per year, and locations in three different areas of the CMU community. Many of the most prominent robotics and artificial intelligence pioneers of the 20th and 21st century have been associated with the Robotics Institute. They include Hans Moravec, Marc Raibert, Raj Reddy, Herbert Simon, Sebastian Thrun, Red Whittaker, and Daniel Wilson.

The Robotics Institute has two primary subsidiaries. The first is the National Robotics Engineering Center (NREC), an organization that allows the institute to commercialize academic discoveries. The NREC maintains strong ties to both the military and industrial communities, and often accepts outside projects based on specific customer needs. This organization makes robotics funding largely self-generating, rather than relying on the school's endowment and budgetary process to keep the Robotics Institute functional. The NREC also engages in substantial outreach to public secondary schools, stimulating interest in the field of robotics and inspiring the next generation of CMU researchers.

The Field Robotics Center (FRC) is an experimental organization that creates robotic prototypes and tests them outside of the laboratory environment. Many of its best-known robots have engaged in external competitions or contracting for specific tasks. Teams from the FRC have figured prominently in the DARPA Grand Challenge races, with the 2005 entries Sandstorm and H1ghlander coming in second and third place, respectively, and the 2007 entry Boss winning the competition.

Foster-Miller Company

350 Second Avenue
Waltham, MA 02451-1196
Phone: 781-684-4000
Fax: 781-890-3489
Website: http://www.qinetiq-na.com/

Foster-Miller Company is a military robotics engineering company that tends to focus on only a small number of niche products but has an excellent reputation for providing extremely rugged, dependable systems. The company was founded in 1956 in Waltham, Massachusetts, where it retains its headquarters despite its acquisition by British military consortium QinetiQ in 2004. Foster-Miller employs approximately 300 highly skilled workers, with a large number of engineers employed in the design, testing, and construction of military robots. It primarily designs robots for hazardous location service, and produces unique composite materials for use in a variety of military and industrial applications.

The company is best known for two military products, the TALON ground robot and the LAST Armor system. The TALON is a tracked ground robot capable of performing a number of hazardous or tedious missions. It has proven popular with American forces in Afghanistan and Iraq. While it is most associated with explosive ordnance disposal, TALON missions have expanded into chemical, gas, temperature, and radiation sensing through the creation of new sensor packages that can be easily swapped onto the robot chassis. More than 2,000 TALONs have been utilized in Afghanistan and Iraq, grossing the company more than $120 million in production contracts.

The LAST Armor system is a series of composite material panels designed to improve the defensive capabilities of Humvees. Unlike other "up-armor" systems, LAST Armor is designed so that an untrained solder can easily remove and replace it without the use of specialized tools and gear. This allows the rapid replacement of damaged panels, which can then be returned for repair and reuse. The system has proven effective at reducing personnel losses due to small arms fire and improvised explosive devices. Recently, Foster-Miller has increased its research into composite materials for use in the construction of unmanned aerial vehicles by other companies, ensuring that it will not become entirely reliant on a single product.

General Atomics

Location: 3550 General Atomics Court
San Diego, CA 92121-1122
Mailing Address: P.O. Box 85608

San Diego, CA 92186-5608
Phone: 858-455-3000
Fax: 858-455-3621
E-mail: PR_Info@ga.com
Website: http://www.ga.com

General Atomics is a defense contractor and nuclear reactor corporation founded in San Diego in 1955. It began as the General Atomic Division of General Dynamics, allowing its parent company to expand into the burgeoning field of nuclear power generation during the first boom of nuclear power plant construction. Today, the company still provides substantial consulting services and design assistance to the nuclear industry, although it is best known for recent forays into the field of unmanned aerial vehicle design. A subsidiary, General Atomics Aeronautical Systems, Inc. (GA-ASI), is responsible for developing the RQ-1/MQ-1 Predator and RQ-9/MQ-9 Reaper UAV systems.

Since 1967, General Atomics has been acquired by a series of energy firms, primarily oil companies. In 1986, brothers Neal and Linden Blue purchased the company, and soon enlisted the services of retired Rear Admiral Thomas J. Cassidy, Jr., to run the company. When GA-ASI was created, Cassidy became its first president, playing a central role in the rise of the firm's defense contracting operations.

The earliest version of the Predator, the RQ-1, made its maiden flight in January 1994. Less than two years later, it entered service with the U.S. Army, Navy, Air Force, and Central Intelligence Agency. In the summer of 1995, Predators were deployed for the first time, performing intelligence, surveillance, and reconnaissance (ISR) missions over the former Yugoslavia. They proved extremely successful due to their excellent sensor packages, long loiter times, slow speeds that allowed extensive surveillance, and relatively unobtrusive auditory, visual, and radar presence.

Predators continued their experimental role in deployments over Serbia and during the enforcement of a coalition no-fly zone over Iraq. In the aftermath of the September 11, 2001, attacks, however, the demand for unmanned systems rose precipitously, such that General Atomics has now produced nearly 400 Predators at a unit cost of less than $5 million, including sensors. Initially, General Atomics envisioned Predators solely as an ISR platform; however, the desire to reduce the delay between spotting a target and attacking it with a manned aircraft led to the development of armed Predators carrying AGM-114 Hellfire missiles. These attack drones, designated MQ-1, met with an immediately enthusiastic response.

Anticipating the military's desire for an even larger, more robust attack UAV, General Atomics undertook a research program using its own resources to develop the Predator B, later renamed the RQ-9 (ISR) or MQ-9 (multirole) Reaper. This system, equipped with a much larger turboprop engine, can carry 15 times more ordnance than its predecessor, yet it costs slightly more than twice as much as its smaller cousin. The Predator C, later called the Avenger, is a jet-powered, stealthy UAV designed to carry at least the same payload as the Reaper but in a much faster airframe. It made its first flight in 2009 but has not yet been formally entered into military service.

GA-ASI continues to invest in the development of larger and more capable UAVs, remaining at the cutting edge of unmanned weapons development. It has also begun to produce sensors designed specifically for its own frames, reducing reliance on external contractors and increasing the corporation's attractiveness to military acquisitions specialists. GA-ASI has also massively increased its production capabilities, growing from two per month in 2003 to more than 10 per month in 2010, for supply to the U.S. military and approved foreign purchasers. GA-ASI has created a lucrative business providing support and pilot training to UAV operators, following the pattern established by Orville and Wilbur Wright's Flying School at Maxwell Airfield. Currently, GA-ASI aircraft are operated by the military forces of Italy, Turkey, and the United Kingdom in addition to the United States.

Honda Motor Company

1-1, 2-Chome
Minami-Aoyama, Minato-ku
Tokyo 107-8556
Japan
Phone: 81-(0)3-3423-1111
Website: http://world.honda.com

The Honda Motor Corporation was founded in 1948, primarily for the production of engines, automobiles, and motorcycles. However, as the company grew, it diversified into a large number of other industrial production fields, including the design of robots. Within the robotic field, Honda has developed two significant subfields: industrial robots to assist in the production of its vehicles, and humanoid robots that have become increasingly lifelike in their movements and ability to replicate human behavior.

After only 26 years of research, Honda's humanoid robots have reached an incredible level of sophistication, personified in their ASIMO (Advanced Step in Innovative Mobility) design. ASIMO has been designed as a helper, which reflects the Japanese cultural proclivity to use robots as companion and assistance machines. ASIMO's technology offers a substantial possibility of not only providing direct assistance to humans but also of creating spin-off technologies to restore mobility to people who have lost use of their limbs. Honda has spent substantial sums researching a Brain Machine Interface, a helmet coated with sensors to detect neural impulses and blood flow within a subject, essentially allowing the ASIMO or other robotic devices to be controlled by thought rather than direct input of commands.

While Honda Motor Company has shown little interest in becoming a major defense contractor, the sophistication of its robots demonstrates the possibilities of robotic technology in the near future. The latest version of ASIMO is capable of bipedal running at six miles per hour, as well as traversing stairways. Even if Honda does not choose to develop military robots, there is ample reason to believe that if ASIMO or a similar humanoid robot becomes available on the consumer market, it will quickly be adapted to military use. Further, the human assist technologies that have been created from the same research are applicable in a military setting, as is the possibility of controlling machines via thought rather than the complicated remote piloting systems in use today.

iRobot Corporation

8 Crosby Drive
Bedford, MA 01730
Phone: 781-430-3000

Fax: 781-430-3001
E-mail: info@irobot.com
Website: http://www.irobot.com

In 1990, Rodney Brooks, Colin Angle, and Helen Greiner founded iRobot, a corporation dedicated to producing consumer robots. iRobot is headquartered in Bedford, Massachusetts, near the Massachusetts Institute of Technology's Artificial Intelligence Laboratory, where all three founders worked. Although the company was not initially founded to specialize in military robots, all three founders had performed various research roles in the creation of military technology. In 1998, iRobot received its first DARPA research contract, beginning an extensive relationship that continues today.

The largest commercial success for iRobot has been the Roomba, a robotic vacuum cleaner capable of autonomous floor cleaning that has sold more than 5 million units. Spin-off variants of the Roomba include the Scooba, designed to wash hard floors, the Looj for cleaning gutters, the Verro swimming pool cleaner, and the Dirt Dog, a heavier-duty version of the original for workshop and industrial applications. The company has also produced toys, including My Real Baby, an animatronic doll first produced in 2000.

Early iRobot designs included Genghis, a small six-legged robot designed to walk by emulating the biological instincts of insects, first built in 1991 and now housed in the Smithsonian Air and Space Museum. Ariel, a biomimetic robot similar to a crab, could be used for counter-mining operations both on land and at sea. When iRobot introduced PackBot in 1998, it quickly drew positive attention from military ground forces, particularly those engaged in explosive ordnance disposal. The PackBot concept was enlarged and expanded with Warrior, iRobot's next-generation, tracked multipurpose robot. The company moved into underwater unmanned vehicles (UUV) with Transphibian, first released in 2008. This small, fin-powered UUV can be used in counter-mining, harbor defense, and underwater surveillance roles.

The company remains heavily involved in the development of new generations of robots for both military and civilian applications, and is at the vanguard of the school of robotics companies that believe small robots engaging in simple behaviors, when linked together, can perform extremely complex activities. Currently, DARPA and iRobot are working on algorithms necessary to guide the individual behaviors of thousands of robots working together in swarms. To support the concept, iRobot has designed LANdroid, a miniature robot weighing less than one pound. Its production cost will be so low that the loss of several LANdroids will be negligible. While each unit moves at only a few miles per hour, a swarm of LANdroids could map an entire large city in a matter of a few hours, given successful programming and sufficient communication links.

DARPA has also funded iRobot's attempts to create a shape-shifting robot made of advanced polymers that can alter it configuration when subjected to electric or magnetic fields. The system utilizes a chemical power source to avoid the rigidity of existing batteries and does not contain any large, solid parts, theoretically allowing it to slither through very small openings. The company expects to create a semisolid machine that can perform a variety of functions while traversing virtually any terrain.

MIT Computer Science and Artificial Intelligence Laboratory

Ray and Maria Stata Center
32 Vassar Street, 32-369
Cambridge, MA 02139-4309
Phone: 617-253-5851
Website: www.csail.mit.edu

The organization currently known as the Massachusetts Institute of Technology Computer Science and Artificial Intelligence Laboratory (CSAIL) was founded in 1959 by John McCarthy and Marvin Minsky. McCarthy soon left to teach at Stanford, where he founded the Stanford Artificial Intelligence Laboratory (SAIL), but Minsky remained at MIT for his entire career. In 1963, the Defense Advanced Research Projects Agency provided a $2 million grant to underwrite programs in artificial intelligence and computation theory at the university. The grant, provided to establish Project MAC (Mathematics and Computing), provided a much-needed boost to what had previously been a collection of pioneering computer programmers with too few resources to pursue all of their ideas. In 1970, Minsky spun off a new organization, the MIT Artificial Intelligence Laboratory, bringing along many of the original Project MAC personnel and creating a renewed focus on the study of artificial intelligence. Other original members formed the Laboratory for Computer Science and turned their attention to operating systems, programming languages, and computer design theories.

In 2003, the disparate organizations merged to form CSAIL under the direction of Rodney Brooks. The current organization allows research groups to form to attack specific problems, with seven areas of primary investigation: artificial intelligence, computational biology, graphics and vision, language and learning, theory of computation, robotics, and systems. The guiding principle of the organization is the notion that improved computation will allow a greater understanding of physical and biological systems, and thus as the capabilities of computers are improved, every field of research endeavor will also benefit. In addition to the research conducted at the CSAIL facilities, the institution has created partnerships throughout industry and academia, and has made the Boston area into one of the critical nodes of robotics research and production, with such companies as iRobot and Foster-Miller in the immediate region.

RAND Corporation

1776 Main Street
P.O. Box 2138
Santa Monica, CA 90407-2138
Phone: 310-393-0411
Fax: 310-393-4818
Website: http://www.rand.org

The RAND Corporation (Research And Development) is a nonprofit defense analysis organization that serves to bridge the gap between theoretical concepts and practical implementation of strategic ideas and new technological innovations. The company is funded through government budgetary support, private endowment, corporate contracts, and universities.

RAND was founded in 1948 in Santa Monica, California, primarily as a think tank in service of the U.S. Air Force. The creation of nuclear weapons had called into question the entire strategic theory of the American military establishment, and the Air Force wished to ensure that an outside agency could provide an objective viewpoint of new ideas being postulated regarding the conduct of future wars. From the early discussions of nuclear policy, RAND branched out into studies of global policy, diplomacy, military affairs, and eventually, virtually every form of government activity.

RAND currently employs approximately 1,700 employees and earns annual revenues of $250 million. These revenues fund further investigations into virtually any topic of interest to the U.S. government and other contracting organizations. RAND publications are often distributed in an unclassified fashion, although the corporation also creates secret reports for government agencies. RAND has a reputation for seeking the top talent within a wide variety of fields and has succeeded to such an extent that more than 30 Nobel Prize laureates have been associated with the company at some point in their careers. RAND has been used extensively to evaluate the American robotic and drone military programs as well as to chart the effectiveness of existing systems and to make recommendations regarding future procurement of robotic systems. These studies, in turn, have helped to define military requirements for future programs and eliminate programs that have proven ineffective.

Stanford Artificial Intelligence Laboratory (SAIL)

Gates Building 1A
353 Serra Mall
Stanford, CA 94305-9010
Website: http://robotics.stanford.edu

The Stanford Artificial Intelligence Laboratory was founded in 1963 by John McCarthy, who had previously cofounded the MIT Computer Science and Artificial Intelligence Laboratory. It initially served as a research node for efforts to develop new programming languages that might make the development of artificial intelligence possible, although its function soon expanded to include robotics, computer science, and information networks. After nearly two decades of semi-independence from the university, SAIL was folded into the university's computer science department, where it remained for more than 20 years. In 2004, Sebastian Thrun assumed the directorship and assisted in the transition away from the confines of departmental bureaucracy, while also leading robotics competition teams to national prominence. The current director of SAIL is Andrew Ng.

SAIL's mission is to "change the way we understand the world," using a multidisciplinary approach to enhance the development of computers, robotics, and artificial intelligence. With its location on the west coast, SAIL is in an excellent position to partner with many Silicon Valley firms, and it has worked on a number of projects with such industry giants as Cisco Systems and Sun Microsystems as well as a host of smaller companies. Under Thrun's leadership, the SAIL entry in the DARPA 2005 Grand Challenge, Stanley, won the race, completing the 132-mile course in under seven hours, thus claiming the $2 million prize. In 2007, the Stanford Racing Team came in second place in the urban version of the Grand Challenge. SAIL researchers are currently engaged in projects involving bioinformatics, cognition, computer perception, game theory, knowledge systems, machine learning, neural networks, and robotics.

Defense Conglomerates

BAE Systems

6 Carlton Gardens
London, SW1Y 5AD
United Kingdom
Phone: 44-1252-37-3232
E-mail: baesystemsinfo@baesystems.com
Website: http://www.baesystems.com

BAE Systems is a company founded through the merger of British Aerospace and Marconi Electronic Systems in 1999. Headquartered in London, but with affiliated companies around the globe, it is one of the largest defense contractors in the world, with 107,000 employees and more than $25 billion in annual revenues. Much of the impetus of the merger came from the need to compete with other defense conglomerates—in the 1990s, a series of major defense mergers threatened to overwhelm the British defense industry. Without this consolidation, British Aerospace might have been purchased or driven out of business by its much larger competitors. The consolidation of American and British firms in turn caused the creation of a continental European conglomerate, EADS, but the British government refused to consider a pan-European merger.

BAE Systems is a moderately diverse contractor; while most of its revenues come from defense contracts, it has avoided the temptation to specialize in only a certain type of military production. It is the largest British manufacturer of aerospace vehicles, military electronics, and naval vessels. The company has been actively involved in the development and production of the F-35 Lightning, the Eurofighter Typhoon, submarines, and aircraft carriers. In the period since its creation, BAE Systems has been heavily interested in acquiring smaller defense companies, causing it to gradually become involved in ground vehicles and smaller weapons systems.

BAE has a reputation for producing extremely sophisticated unmanned systems. Several of their platforms are stealthy UAVs capable of unarmed reconnaissance or deep strikes in a hostile environment. In particular, the MANTIS (Multispectral Adaptive Networked Tactical Imaging System) has drawn significant attention because it has one of the largest payload capacities of any unmanned combat aerial vehicle, and its radar signature, electronic countermeasures, and autonomous capabilities make it capable of operating against even the most well-designed air defense networks.

Unlike many of its peer competitors, BAE Systems is an almost purely military contractor—over 95 percent of its revenues are drawn from military sales. Despite being a British company, its largest client is the U.S. Department of Defense. Like virtually all huge contractors, significant allegations of political corruption, particularly as related to international arms sales, have dogged the company for its entire existence. However, such accusations have not hindered its growth into one of the largest defense manufacturers in the world.

Boeing

100 North Riverside
Chicago, IL 60606
Phone: 312-544-2000
Website: http://www.boeing.com

Boeing is one of the largest aerospace and defense firms in the world, and also one of the oldest continuously operating airplane manufacturers. It was founded by William E. Boeing in Seattle in 1916 and grew steadily until its merger with McDonnell Douglas Aircraft Corporation in 1997. Currently, Boeing's corporate headquarters are in Chicago, Illinois, although it has facilities in numerous locations throughout the United States. Boeing employs 164,000 individuals and has annual revenues of $64 billion.

Boeing was initially formed as an airplane construction company, focusing on the niche of seaplanes. During World War I, the company manufactured airplanes for the U.S. government, but in the immediate postwar period, the company nearly collapsed due to a flood of surplus military airplanes onto the civilian market. Boeing survived primarily though diversification—in the decade after the war, the company produced many types of consumer goods other than airplanes, waiting for market demand to resume. By the 1930s, American military interest in aviation had resumed, and the company created major programs in bomber and fighter aircraft for the War Department. In the same decade, Boeing displayed a caution about becoming too dependent on a single revenue stream, and even though it became increasingly dedicated to aviation, it did not remain confined to military production. As such, the company was able to pioneer the field of passenger transportation, essentially creating a market to serve its own production capabilities.

During World War II, the company's primary focus shifted to bomber aircraft, most notably the B-17 and B-29 bombers that devastated Germany and Japan, respectively. In the aftermath of World War II, having learned from the previous postwar experience, Boeing remained involved in military production but again diversified well beyond aircraft. Boeing was one of the first companies to become involved in the production of intercontinental ballistic missiles for the U.S. military. At the same time, Boeing expanded its holdings in commercial aircraft and explored opportunities to serve as a component subcontractor in other major military systems.

Boeing is largely regarded as one of the great success stories in American manufacturing, and the company continues to produce high-quality aviation equipment while also pursuing new fields of production. Naturally, seeing the military shift toward unmanned vehicles, the leadership of the company made aggressive moves to enter the UAV market. One of the chief ways that Boeing has succeeded in this new field is the acquisition of specialty companies, such as Insitu, one of its main subsidiaries. Through such purchases and strategic partnerships, Boeing has been able to produce the ScanEagle, NightEagle, Integrator, and Interceptor platforms, as well as a host of less well-known and experimental aircraft.

European Aeronautic Defence and Space Company

Mendelweg 30
2333 CS Leyde
Netherlands
Phone: 31-71-524-56-00
Website: www.eads.com

The European Aeronautic Defence and Space Company (EADS) is a global defense conglomerate that specializes in high-technology military hardware. It is headquartered in Leiden, Netherlands, and employs more than 120,000 workers. It was founded in 2000 as a merger of three national military companies: France's Aérospatial-Matra, Germany's DaimlerChrysler Aerospace, and Spain's Construcciones Aeronáuticas SA. Its subsidiaries include Airbus, Astrium, Cassidian, and Eurocopter, along with a host of smaller, more specialized companies. One of the largest defense contractors in the world, EADS was formed largely in response to a pair of mergers in the United States, that of Lockheed Martin and Boeing/McDonnell Douglas. To remain competitive in military contracting, the European firms chose to combine their resources and streamline their design and production of new hardware. While its traditional markets have primarily been in Europe and Africa, EADS now does a majority of its contracting with the U.S. military.

EADS produces a number of unmanned vehicles, including the Eagle, a French long-range ISR platform that is comparable to the RQ-4 Global Hawk. EADS also produces a line of unmanned helicopters, largely through its Eurocopter subsidiary. For the German military, EADS is the lead contractor for the unmanned combat aerial vehicle the Barracuda. This stealthy, jet-powered strike aircraft is one of the most advanced UAVs in the world, demonstrating the company's capabilities. EADS has also engaged in a number of strategic partnerships, including a sub-mission under Northrop Grumman's lead to build the U.S. Air Force's next line of refueling aircraft, the KC-45. When the bidding process was reopened, Northrop Grumman withdrew from the competition, but EADS chose to present a separate design based on the Airbus 330. EADS continues to aggressively pursue acquisitions of specialized high-technology design and engineering firms, and will likely remain one of the largest defense contractors in the world for the foreseeable future.

Lockheed Martin Corporation

6801 Rockledge Drive
Bethesda, MD 20817
Phone: 301-897-6000
Website: http://www.lockheedmartin.com

Lockheed Martin is one of the largest military production companies in the world, with headquarters in Bethesda, Maryland; more than 123,000 employees; and annual revenues of more than $45 billion. The company was formed by the 1995 merger of Lockheed Corporation and Martin Marietta, two companies that had previously competed in the field of aeronautics. While not exclusively dedicated to military contracting, nearly 75 percent of the company's revenues come from military sales. In 2008 alone, Lockheed Martin won the rights to more than $36 billion in government contracts, and over 7 percent of the Pentagon's 2009 budget went to the company. The company is primarily involved in aeronautics, electrical systems, information systems, and space platforms. Thus, the company has developed a reputation for focusing on extremely high-technology innovations, with less focus on more mundane munitions. Key Lockheed Martin systems that are still a part of the American military arsenal include the Trident missile, F-16 Falcon, F-22 Raptor, C-130 Hercules, and Titan rockets.

In 1998, Lockheed Martin and Northrop Grumman considered a merger that would have created the largest military contracting company in world history, but U.S. government regulators expressed concerns that such an enormous company might create unacceptable disruption within the military procurement system and would also potentially be fraught with opportunities for misconduct. Since Lockheed Martin's foundation, it has paid nearly $600 million to settle over 50 cases of contractor misconduct. In 2011, the company's computer network was penetrated by a cyber attack that penetrated sensitive materials, although the exact losses have not been publicly disclosed.

Northrop Grumman Corporation

1840 Century Park East
Los Angeles, CA 90067-2199
Phone: 310-553-6262
Website: http://www.northropgrumman.com

Northrop Grumman is one of the largest defense contracting companies in the world, with annual revenues of $15 billion and a workforce of more than 72,000. It was formed in 1994 when the Northrop Corporation bought the Grumman Aerospace Corporation. Prior to the merger, each company had a long and proud history of military industrial production, with each founded in the 1920s and making major contributions to the Allied war effort in World War II. Unlike many of the other enormous military producers, Northrop Grumman tends to focus on the final assembly of the largest and most complex military vehicles in the world. It is the largest naval vessel constructor in the world and the manufacturer of every U.S. Navy aircraft carrier currently under contract. The company is also responsible for the B-2 Spirit bomber aircraft, the RQ-4 Global Hawk, and RQ-5 Hunter platforms.

Although Northrop Grumman is associated with some of the best-known individual aircraft and naval vessels, it maintains a broad base in smaller and supporting systems manufacturing as well as a heavy interest in sensor systems for both manned and unmanned platforms. One of its major subsidiaries, Remotec, specializes in the production of explosive ordnance disposal robots and remotely piloted vehicles created for hazardous materials applications.

As the sole source for many of the highest-value systems in the American military arsenal, Northrop Grumman has a certain ability to act with impunity and can continue to expect future major contracts even if there are cost overruns or irregularities in current production cycles. The company has survived a series of major scandals, and settled cases with the U.S. government regarding overcharges, supplying faulty equipment, and violations of the International Traffic in Arms Regulations, including the transfer of highly sensitive computer guidance systems to several nations. While such scandals illustrate the need for diversification of producers and suppliers in the national military production chain, even detractors of the company must acknowledge that it is one of the largest and most technically competent military contractors currently in operation, and it has proven repeatedly that it has the ability to deliver extremely complex hardware that would potentially be beyond the capabilities of even its largest competitors.

QinetiQ North America

7918 Jones Branch Drive, Suite 350
McLean, VA 22102
Phone: 703-752-9595
Website: http://www.qinetiq-na.com

QinetiQ (pronounced "kinetic") is a high-technology military engineering contractor consortium that has recently become extremely active in the fields of military robotics and unmanned aerial vehicles. It was founded in 2001 after the collapse of the United Kingdom Defence Evaluation and Research Agency, with QinetiQ absorbing the largest splinter of the previous organization. The company is headquartered in Farnborough, United Kingdom, but it has locations in the United Kingdom, United States, and Australia, employing more than 14,000 individuals and earning more than $1.7 billion in annual revenues.

In 2004, QinetiQ acquired Foster-Miller, and with it, the Foster-Miller TALON ground combat robot. This acquisition has proven extremely lucrative, as more than 2,000 TALONs have been deployed to Iraq and Afghanistan, primarily for use in explosive ordnance disposal. The TALON has proven extremely rugged, though not bomb-proof, and the company has also established a significant revenue stream from repairing and reconditioning robots damaged in the line of duty, going so far as to establish in-theater "robot hospitals" to speed the turnaround time for repairs. From 2006 until 2008, former Central Intelligence Agency director George Tenet ran the company, bringing a large amount of business through his excellent contacts in the American political and military leadership.

Currently, QinetiQ is moving strongly into the unmanned aerial vehicle field. The QinetiQ Zephyr set a record for endurance by a powered UAV by remaining aloft for more than 14 days. The Zephyr uses solar cells to charge batteries during the day and then operates through each night on the battery power. While its payload is extremely small due to the need to keep the airframe light, the system's endurance makes it an attractive option for long-term observation.

Raytheon Company

870 Winter Street
Waltham, MA 02451-1449
Phone: 781-522-3000
Website: http://www.raytheon.com

The Raytheon Company is one of the oldest continually operating defense contractors in the United States. It is also one of the largest such corporations, with annual revenues in excess of $25 billion and a workforce of more than 70,000 individuals. Raytheon specializes in weapons and military electronics, including special-mission aircraft. Raytheon is also by far the largest producer of guided missiles in the world.

The company was founded in 1922 in Cambridge, Massachusetts, by Vannevar Bush, Laurence K. Marshall, and Charles G. Smith. Its current headquarters are in Waltham, Massachusetts, although it has subsidiaries and offices throughout the world. Although the company, founded as the American Appliance Company, originally focused on refrigeration technology, its focus soon shifted to the emerging field of electronics, and by World War II, it became a major defense contractor specializing in the production of magnetrons. These devices are the vital component in radar systems, and by the end of the war, Raytheon had produced 80 percent of the wartime magnetrons. Shortly after the war, the company introduced the first microwave ovens, using the same magnetrons that had previously detected enemy aircraft.

In 1948, Raytheon produced its first guided missiles, opening an extremely lucrative field of production that the company has dominated for the subsequent six decades. Raytheon produced a wide variety of missile types during the Cold War, including the Lark, Sparrow, Hawk, Phoenix, and Patriot models. The Patriot gained substantial attention during the 1991 Persian Gulf War, producing a large number of defense contracts for the company and provoking an increased specialization in defense manufacturing. Shortly after the war ended, the company began divesting itself of holdings in nondefense sectors, choosing to specialize even further and trusting that military contracting would produce substantial opportunities for profit. The decision has resulted in the creation of one of the five largest defense contractors in the world, and responsibility for designing and producing the most sophisticated missiles and missile defense systems in the American arsenal. Raytheon continues to produce and upgrade the MIM-104 Patriot air defense missile, as well as the BGM-109 Tomahawk cruise missile and the AGM-129 Advanced Cruise Missile, each of which saw substantial usage in Iraq from 2003 to 2012. While the company does not typically assume the lead role for most robotic and drone systems, it does produce components to enhance the systems and remains a vital supplier of military technology for the United States.

8

Print and Nonprint Resources

B ecause the field of military robotics is relatively new, there is a decidedly small literature that deals with the fundamental issues presented by the use of remotely piloted vehicles, military robots, and autonomous weapons on the battlefield. What follows is a guide to some of the best introductory works on robotics, artificial intelligence (AI), and unmanned military vehicles. Many of the best resources are available online; it is unsurprising that such a technologically oriented and novel field would generate substantial digital resources. The sources included in this chapter are selected either because they are widely available, or they are the best works on their subject, or ideally, they are both useful and obtainable. Many of these works contain extensive bibliographies in their own right.

Books and Articles

Andriole, Stephen J., ed. 1985. *Applications in Artificial Intelligence*. **Princeton, NJ: Petrocelli Books.**

This work is a collection of articles showing the state of the art in the field of artificial intelligence in 1985. It is a useful snapshot to create a comparative sample to see the massive progress of the past three decades. In many cases, the reality far outstripped the predictions of experts in the field, even when they found their own ideas to be far-fetched, and the development of artificial intelligence continues to gather speed. Even in 1985, Andriole, the former director of the Defense Advanced Research Projects Agency Cybernetics Technology Office, found that most AI funding came from the Department of Defense. Collectively, the authors envisioned the development of autonomous unmanned aerial vehicles (UAVs), although they did not predict precisely how such devices would be created and used. While predictions about the future of technology and its applications in conflicts are somewhat dubious, this work is instructive in how accurate predictions sometimes prove to be, when made by subject matter experts in their own field.

Arkin, Ronald C. 2009. *Governing Lethal Behavior in Autonomous Robots.* **Boca Raton, FL: Taylor & Francis Group.**

Arkin's work serves not only as a guide to the technical aspects of system design for autonomous robots, but also an examination of the moral, philosophical, and legal aspects associated with the use of autonomous robots on the battlefield. He presents a clear distinction between robots and drones operating under human supervision, called remotely piloted vehicles (RPVs) and those utilizing programming to determine the correct application of force, autonomous machines. Arkin believes it is a major change in the nature of robotic design to authorize lethal force in programming, and one that might open many potential dangers if used incorrectly or too frequently in the modern combat environment.

Arkin believes that artificial intelligence capable of ethical behavior is a vital precursor to the authorization of lethal force in autonomous machines. While robots might make war more humane by limiting civilian casualties through adherence to the international laws of war, this advance will be fraught with peril until substantial advances in programming have been achieved. According to Arkin, researchers must determine how to translate international laws governing battlefield ethics into programming that can utilize real-time reasoning. Robotic behavior must be controlled by rigorously defined ethical boundaries. Sensory capabilities and target discrimination need massive improvements before robots with lethal capabilities can be allowed to roam the battlefields of the future.

Armitage, Michael. 1988. *Unmanned Aircraft*. **London: Brassey's Defence Publishers.**

While this work is thoroughly dated in regards to aircraft that it considered state-of-the-art in 1988, it is still useful because it contains an excellent history of early drones and guided aircraft. To the author, the definitions of drones, unmanned aircraft, and cruise missiles differ from the modern understanding, but once the modern terms are substituted, there is much to be admired in this work. Armitage reminds the reader that unmanned aircraft have a long history of inflicting casualties in both the military ranks and the civilian populace. Even works that acknowledge the early history of UAVs often overlook the period from World War II to the Vietnam War, and in this section, Armitage truly hits his stride.

Armitage considers cruise missiles to be a class of drones, which does not match the modern usage, but is a valuable technological development path to follow. He sees the German V-1 as the inspiration for the American development of cruise missiles and spends a substantial amount of time discussing the earliest cruise missile platforms, including the Snark, the Navajo, and the Matador. In particular, he points out that in the first half of the 1950s, while intercontinental ballistic missiles received $26 million in funding, the cruise missile programs received $450 million. The cruise missile continued to be a vital part of the American, and later Soviet, arsenals, in part due to its versatility and the fact that this class of weaponry tended to be ignored by virtually every arms limitation treaty.

Brooks, Rodney. 2002. *Flesh and Machines: How Robots Will Change Us*. **New York: Pantheon Books.**

Brooks, a pioneer in the development of autonomous robots and artificial intelligence, argues that humans will incorporate robotics into their everyday lives, including inside our own bodies. He envisions a blending of natural and artificial resources in the human body, enhancing the physical and mental capabilities of individuals. This blending of flesh and machine will prevent the "robotic takeover" popularized in science fiction because humans will essentially become the machines, maintaining control by assuming the mantle of the evolved organism augmented by science. Brooks foresees a major break with the past, which he expects to occur around 2020. Although Brooks has been at the forefront of robotics research for more than three decades, he is somewhat naïve regarding the likelihood that human augmentation will be utilized for military purposes, tending to predict the best possible outcomes with little regard for past human behaviors regarding new technology and its employment in war.

Brooks, Rodney A. and Anita M. Flynn. 1989. "Fast, Cheap and Out of Control: A Robot Invasion of the Solar System." *Journal of the British Interplanetary Society* **42, 478–485.**

This article proposes a radically different approach to robotic design and extraplanetary exploration. Specifically, rather than the previously dominant approach of incredibly complex, and correspondingly expensive, robotic systems that remained under the control of operators on Earth, Brooks and Flynn call for small, cheap, mass-produced robotic explorers. In their view, the new method would reduce the time lag from mission conception to implementation, would reduce the launch mass of each mission, and would eliminate the communications problems that had hindered earlier efforts. Although the reliability of each individual robot might be reduced, the overall probability of mission success would increase due to the redundancy of robots, each of which could be considered expendable.

This article, and its underlying argument, met with tremendous resistance in the scientific community, particularly at the National Aeronautics and Space Administration (NASA), where researchers often waited for years to be included in an exploration mission, and tended to pack exploration vehicles with heavy, power-consuming equipment and complex programming. Ironically, they also insisted on terrestrial control, which led to navigation errors triggered by time lag and robotic shutdowns when communication could not be maintained. However, Brooks was able to demonstrate the utility of the idea when two of his researchers at the Massachusetts Institute of Technology (MIT) Mobile Robot Group designed and produced a one-kilogram robot, Genghis, in under 12 weeks. The device, which moved at up to three kilometers per hour, could climb over meter-high rocks and could be outfitted with a variety of exploration devices. Brooks and Flynn also proposed the creation of a host of small robots to bring samples back to a larger, more complex machine, which would act as a scientific lab, power charging station, and communication system.

The ideas in this article remain easily transferrable to the modern battlefield environment. Rather than relying on an expensive, complex logic system in a large vehicle, military roboticists have begun to pursue the idea of tiny, autonomous systems engaged in intelligence-gathering activities. Like the exploration mission imagined in this article, it is conceivable that a robotic "mother ship" might oversee the autonomous activity of hundreds of smaller vehicles.

Clark, Richard M. 2000. *Uninhabited Combat Aerial Vehicles: Airpower by the People, for the People, but Not with the People.* **CADRE Paper No. 8, Maxwell Air Force Base, AL: Air University Press.**

This work presents an unconventional interpretation that drones, remotely piloted vehicles, and UAVs are all interchangeable concepts. Clark argues that kites were the first UAVs, used primarily for psychological operations, geometric calculations useful for besiegers, and communication. This makes for an interesting view of the history of military kites, and Clark's work remains strong when discussing the early attempts to create unmanned aircraft to carry explosives, such as Elmer Sperry's Flying Bomb project and Charles Kettering's Bug. Clark also presents a fascinating examination of Project Aphrodite, an attempt to utilize obsolete bomber aircraft as flying teleoperated bombs.

The heart of the work studies the use of drones and UAVs in the modern era, beginning with the Vietnam experience of the Firebee, Fire Fly, and Lightning Bug models. Clark's examination of the use of UAVs in the Balkan region, where General Atomics products such as the Gnat and Predator made their first combat flights, draws on personal experiences and official sources to paint one of the most complete portraits of UAV usage in the 1990s that is currently available. While Clark envisions UAVs as a potentially perfect mechanism to suppress enemy air defenses in the future, he cautions that humans must remain in control of weapons release, lest the military accidentally create an autonomous weapon capable of lethal decision making.

Drew, John G., Russell Sharer, Kristin F. Lynch, Mahyar A. Amouzegar, and Don Snyder. 2005. *Unmanned Aerial Vehicle End-to-End Support Considerations.* **Santa Monica, CA: RAND.**

The authors, writing on behalf of a RAND corporation study, examine some of the financial and other considerations that are often overlooked when decision makers are comparing unmanned and manned vehicle platforms. They also compare the original specifications in the design plans of the RQ-1 Predator, RQ-4 Global Hawk, and RQ-9 Reaper and compare them to the finished product of each aircraft. Whereas the manned aircraft industry has had decades to improve on the design and manufacturing process, unmanned aircraft are still in a relatively new stage, and the ability to mass-produce most UAVs and robots did not exist in 2005. The authors note that General Atomics, producer of Predators and Reapers, was still handbuilding aircraft at a single facility as late as 2005, leading to a major chokepoint in the production process.

The authors make substantial comparisons between UAVs and their alternatives, sometimes with surprising results. They note, for example, that the Global Hawk has a wingspan broader than a Boeing 737, a size that greatly limits the potential airfields from which it can operate. Further, despite its large size, its sensor package cannot use both the infrared and electro-optical spectrum at the same time, as these modes shared optics. Even though the Global Hawk has a massive loiter time, making it a better persistent surveillance platform than a satellite, the remote locations this type of vehicle was being used in meant that at least three Global Hawks were required to maintain continual coverage of an area. Thus, many of the assumptions about the utility of these platforms proved to be in error, particularly in regard to cost savings and area coverage. The authors conclude with a nice comparison of the UAVs in production to those in development in 2005, allowing the reader to see the extremely rapid rate of change prevalent within the UAV field at the beginning of the 21st century.

George, F. H. 1986. *Artificial Intelligence: Its Philosophy and Neural Context.* **New York: Gordon and Breach.**

This work examines cybernetics as a means to approach artificial intelligence and to model aspects of complex, real systems of cognition. George spends a substantial amount of time describing the various architectures created for logic systems. He argues that the recall of stored material is a necessary aspect of intelligent behavior, something that computers already do far more effectively than humans. However, the basic problem of cognition, to George, is the ability to recognize patterns, something that humans do instinctively but that computers struggle to perform except in specialized systems. This work has little discussion applicable specifically to military robotics, but it is an excellent introduction to the underlying concepts and history of artificial systems and computer logic, for any reader desiring a fairly technical explanation of logic systems.

Griswold, Mary E. 2008. *Spectrum Management: Key to the Future of Unmanned Aircraft Systems?* **Maxwell Paper No. 44. Maxwell Air Force Base, AL: Air University Press.**

This work argues that bandwidth and broadcast spectrum are the key challenges to future UAV development and utilization because current generation UAVs require so much of both that the airwaves have become clogged. Griswold estimates that by the year 2015, the sensor packages on UAVs will have become so sophisticated that they will require a prohibitively high amount of data transfer capability. She notes that during Desert Storm in 1991, Department of Defense satellites were insufficient to provide the required bandwidth for military operations during peak usage, requiring the lease of NATO and commercial satellite capabilities. By the time of Operation Allied Force, the bandwidth requirements had doubled, and in the initial stage of Operation Enduring Freedom, the bandwidth requirement was eight times higher than in Desert Storm, despite having a force only one tenth the size of the previous deployment. The Global Hawk program alone uses five times as much bandwidth as the entire Desert Storm communication system. Griswold concludes the work by arguing that optical data links using laser communications might offer a potential solution, as they can carry greater bandwidth with less chance of interception, jamming, or interference.

However, to make such a system operable will require a substantial research and development commitment that has thus far been lacking.

Gutkind, Lee. 2006. *Almost Human: Making Robots Think*. New York: W. W. Norton.

Gutkind spent four years observing the Robotics Institute at Carnegie Mellon University in Pittsburgh, Pennsylvania. He identified a major dichotomy within the robotics discipline, examining the balance between code-writers, who spend most of their time developing programs and software to determine the behavior of robots, and fabricators, whose primary interest and effort focuses on the actual construction of robots. According to Gutkind, true breakthroughs in robotics require the ability to think deeply about the nature of life, both organic and artificial, and the resources to blend programming and engineering skills.

This work primarily examines robotics from a programming standpoint, describing the multitude of approaches to the question of how robotic thought should be designed. Answers range from preprogrammed reactions when robots are confronted with specific stimuli to group processing by individual robots. To Gutkind, the ultimate goal is the creation of a neural network capable of emulating the connections and abstract reasoning of human brains. Gutkind's view of robotics demonstrates that, as in any relatively new field of study, while failure is far more common than success, the indomitable will of leaders in the field keeps the discipline moving forward. In spite of the failures, the capabilities of modern robots far outstrip those of even the recent past, with new developments occurring in an exponential fashion.

Perhaps the most useful aspect of this work is the outside perspective that it brings to the examination. Robotics, like many cutting-edge technologies, attracts its share of zealots and egotists, some of whom promote the future of their discipline so aggressively that they lose sight of the current state of affairs. Because Gutkind is not involved with the development of robots, he has a certain amount of distance that he can apply to the subject. His first-person descriptions of individual robot field tests present a snapshot into the real-world applications of modern robots.

Haugeland, John. 1985. *Artificial Intelligence: The Very Idea.* **Cambridge, MA: MIT Press.**

This work considers the question of what artificial intelligence truly is and whether popular and scientific assumptions about the nature of cognition are truly accurate. Haugeland commences with an examination of classical conceptions of the mind, finding such a consideration to be a product of the Rennaissance. Thomas Hobbes saw thinking as a mental discourse, following methodical rules and leading to a rational product. To Hobbes, there was no distinction between thoughts and writing beyond that one was internal and the other external to the body. René Descartes divorced thought, which he considered symbolic, from objective reality, thus creating the concept of the modern mind, and establishing the idea of a mental and physical duality, by which physical laws and reason need not follow the same patterns. David Hume wanted to discern the laws of the mind, as he believed that individual thoughts involved the movement of small matter within the brain.

Essentially, Haugeland's work is a history of the pursuit of artificial intelligence, with forays into the concept of what "intelligence" actually denotes. To Haugeland, intelligence is the ability to solve problems, and solutions are found via heuristically guided searches. He is not convinced that true artificial intelligence is impossible, even though many of his contemporaries argued that it was so far into the future as to be unworthy of consideration. In many ways, the work is most useful for comparison to 2012, as it shows the enormous number of hurdles that have been overcome even if true artificial intelligence has still not been created.

Holder, Bill. 2001. *Unmanned Air Vehicles: An Illustrated Study of UAVs.* **Atglen, PA: Schiffer Publishing.**

Holder's work is one of the few examinations of unmanned aerial vehicles that does not focus solely on the United States or a small sample of UAV developers and users. Instead, he adopts a much broader approach to the issue of the history of remotely piloted vehicles in the world and provides a snapshot of the state of UAV usage in 2001 with a major section on other early adopters of the technology. He makes solid links between the iterations of unmanned aircraft, showing the relationships between the earliest guided missiles and the latest cruise missiles. He also makes an interesting argument for the repurposing of obsolete manned aircraft, which might be given new life by removing the life-support and protection components, substituting a remote sensing system, and thus creating remotely piloted combat aircraft from existing airframes.

Kenyon, Henry S. 2008. "U.S. Robots Surge onto the Battlefield," *Signal* **(March). http://www.afcea.org/signal/articles/templates/ Signal_Article_Template.asp?articleid=1523&zoneid=228.**

This article offers a brief profile of the Robotic Systems Joint Project Office (RSJPO), demonstrating the types of robotic systems being used in 2008 in Iraq and Afghanistan. The author finds that a robotic surge into Iraq was having an effect, and was politically less costly than sending more troops to the region. In 2004, the Department of Defense requested 162 robots from RSJPO. The following year, the number increased tenfold, and by 2008, more than 5,000 robots were in Iraq and Afghanistan. To support this surge, RSJPO opened a Joint Robotics Repair Facility in both Iraq and Afghanistan, with the ability to repair or replace robots within four hours of receiving them at the facility.

Krishnan, Armin. 2009. *Killer Robots: Legality and Ethicality of Autonomous Weapons.* **Burlington, VT: Ashgate.**

Krishnan's work seeks to remind the reader that the greatest challenges of autonomous weapons may not come from the technological problems associated with their creation, but rather from the host of legal and ethical implications that arise when one contemplates allowing a machine to make life-and-death decisions on the battlefield. Already, unmanned vehicles and robotic platforms are displaying a remarkable level of autonomous behavior, and many of the current prototype designs take that autonomy even further. With the development of swarming robots and armed nanobots, and the possibility of true artificially intelligent machines, the fundamental nature of conflict may soon change.

Krishnan sees international law as one way to potentially regulate the development and deployment of "killer robots," although he acknowledges that in the past, international agreements have not always kept new technologies from proliferating. Of course, one of the fundamental issues that Krishnan raises, the ethics of warfare, includes the implicit problem that a single standard of ethical behavior in conflict does not exist for the global community. However, he believes that there are certain general agreements, even if the details of every situation are not settled, and that the guiding principles of previous ethical behavior should allow at least the creation of certain boundaries governing acceptable use of autonomous weaponry.

Lehner, Paul E. 1989. *Artificial Intelligence and National Defense: Opportunity and Challenge.* **Blue Ridge Summit, PA: TAB Books.**

Lehner's book examined the status of artificial intelligence research and the potential applications of interest to the American military. One of the primary topics of his work is the proposed Strategic Defense Initiative (SDI) system, which was designed to counter enemy intercontinental ballistic missiles. At the time he wrote the book, SDI researchers were pursuing many different approaches to the question of how to intercept and destroy an inbound missile. Lehner argued that the SDI system would function only if it was created as an autonomous system capable of detecting and countering threats without waiting for the delays required by human control of the system. He noted that such a system would require an automatic programming technique, including an estimated 10 million to 100 million lines of code, any one of which could trigger a failure in the system if it was miswritten.

Lehner also noted that the Defense Advanced Research Projects Agency (DARPA) had sponsored development of advanced artificial intelligence research, robotics, and a strategic computing program, all of which could be expected to bear fruit in the following decades. He concluded the work by predicting, correctly, that "most of the national defense applications of AI presently [1989] being pursued will not succeed in the near-term development of operationally useful systems, despite the fact that many of those programs have the specific objective of developing operationally functional systems in the near future" (187). While he was correct in the assertion that then-current programs failed to produce the results DARPA hoped to achieve, the programs created a network of relationships between the Department of Defense and key research institutions. They also provoked a massive research effort to develop a fundamental understanding of the principles of robotics, rapidly accelerating the development of robotic and remotely piloted vehicles.

Martin, Matt J. and Charles W. Sasser. 2010. *Predator: The Remote-Control Air War Over Iraq and Afghanistan: A Pilot's Story.* **Minneapolis, MN: Zenith Press.**

Martin and Sasser present a first-person account of a relatively early adopter of the armed unmanned aircraft technology. One of the authors, Martin, spent nearly four years flying the RQ-1 and MQ-1 Predator over Iraq and Afghanistan, in missions ranging from intelligence-gathering and surveillance to armed strikes on Al Qaeda militants and providing close air support to U.S. Marines in Fallujah. It is striking in that Martin had to change his own preconceptions of the aircraft and its capabilities—as a pilot, he went through a major transition from flying inside his aircraft to controlling it from a ground station located thousands of miles away. Martin's transformation, from skeptic to true believer, mirrors that of much of the U.S. military and intelligence community.

Martin exposes many of the pitfalls of the technology, including the unexpected errors that inevitably arrive with any new weapon platform. He points out that using rated pilots to remotely fly the aircraft has led to a much greater ability to adapt to unexpected changes in the environment around the aircraft, as they are capable of instinctual corrections that might not occur with individuals who never earned a pilot's license. In this viewpoint, Martin paralleled the stance of the U.S. Air Force, which adamantly refused to allow any nonrated personnel to become remote pilots, a fact that has reduced the available number of operators but also contributed to a lower accident rate than their fellow services' operators.

Over his time as an RPV pilot, Martin became extremely reverent of the technology he was allowed to employ, which enabled him to demonstrate the different missions that the platforms were able to perform. In particular, he is proud of the quick target-to-fire decision process, allowing an almost instant response to glimpsing a high-value target. He continually discusses the importance of finding and eliminating Abu Musab al Zarqawi, the commander of Al Qaeda in Iraq, such that it almost becomes an obsession, but he also contradicts many of the public assumptions about the war, including the notion that collateral damage is not considered when launching drone strikes.

Martin illustrates some of his frustrations with the current design of the UAVs in operation over Iraq and Afghanistan, including a counterintuitive and clunky system for actually flying the aircraft, and the irritating delay that can occur when issuing commands to the UAV. He also demonstrates some of the unique stresses that come with serving as a remote pilot, including an eloquent discussion of the dangers of posttraumatic stress disorder (PTSD) among pilots. As he shows, unlike their manned aircraft counterparts, UAV pilots are required not only to engage targets, but then to loiter over them, surveying the damage done by their attacks. Thus, the Predator pilot is required to see the devastation that can be caused by a Hellfire missile striking a vehicle full of combatants, or worse, the horrors of collateral damage if an innocent civilian moves into the path of munitions. The first-person account explains the problems associated with fighting a war during the day and then attempting to return to a normal family life in the evening. Because almost no other first-person accounts have been produced by UAV pilots to date, this work is unique, and thankfully, it is a well-written exemplar of military autobiography.

McDaid, Hugh and David Oliver. 1997. *Smart Weapons: Top Secret History of Remote Controlled Airborne Weapons*. New York: Welcome Rain.

This work is a basic history of remotely piloted vehicles, with excellent pictures to illustrate the various airframes in operation at the end of the 20th century. Perhaps most interesting and useful in the work is the inclusion of a collection of prototypes and cancelled aircraft that are normally ignored by similar works. The book contains an impressive number of images and technical details, although the analysis is fairly light and mostly speculative, rather than grounded in applications.

Nocks, Lisa. 2007. *Robot: The Life Story of a Technology.* **Westport, CT; Greenwood Press.**

Nocks has created an excellent overview of the development of robots, making this a top choice for a quick history of the field. She places special emphasis on the idea that humanoid configurations should be created for the convenience of their human counterparts—because human cities, for example, are inherently designed for human locomotion, robots that emulate human physical behavior will immediately have access to human environments. From a military standpoint, Nocks's argument is that humanoid-shaped robots would be able to assume the most dangerous military tasks that humans normally perform, such as the need to move through an urban area or other dangerous terrain held by the enemy. There is not much focus on robotic warfare in this work, but the broad examination of robotic design makes it a worthy starting point for examinations of military applications of robots.

Perkowitz, Sidney. 2004. *Digital People: From Bionic Humans to Androids.* **Washington, DC: Joseph Henry Press.**

Perkowitz sees technology progressing inexorably toward the development of true androids, and envisions the creation of a "companion race" that complements humanity but has different capabilities, functions, and values. He begins with an examination of the most common portrayals of artificial beings in literature and film, as he believes the common cultural understanding of artificial life will drive the creation and utilization of such robotic systems when they actually become possible.

Perkowitz lingers on the specific hurdles preventing the construction of androids, including program storage, sensory capabilities, and appearance. He considers the problems of integrating the machine mind and body, and how to simulate such human physical characteristics as movement and facial expressions. While he recognizes that there are currently few compelling reasons for pursuing the development of androids versus other potential forms of robots, the efforts still continue. The author finds such efforts grounded in arrogance and illogic, unless the actual goal is the creation of the companion race he envisions. While he is correct that there are fundamental challenges standing in the way of androids, in the less than a decade since the publication of his work, robotics researchers have made enormous strides in confronting each of the issues he examines, such that the notion of companion robots has almost become a reality in certain areas of the world.

Rosenberg, Jerry M. 1986. *Dictionary of Artificial Intelligence and Robotics.* **New York: John Wiley & Sons.**

While Rosenberg's work is now a bit dated, and of necessity does not incorporate more than two decades of new robotic developments, this work did establish the fundamental definitions of many robotics terms, clarifying the terminology and providing a shared lexicon for researchers in this field. It contains very short, straightforward, and useful definitions placed in context. Perhaps most importantly, it defines a robot as a machine that contains sensing instruments and has an ability to respond to its environment via programming; or a machine that can perform specific tasks automatically. The lack of cross-references hinders the usefulness of the book somewhat, but it remains a handy guide for the study of the field.

Singer, P. W. 2009. *Wired for War.* **New York: Penguin.**

Singer makes the case that the incorporation of robotics into the modern battlefield constitutes a revolution in military affairs, and thus a decisive change in virtually every aspect of how wars are started, fought, and completed. In his view, the robotics revolution is every bit as significant as the advent of nuclear weaponry and has the potential to trigger just as many social changes as the development of the atomic bomb. His is a broad examination of the current state of the field of military robotics, based largely on personal experiences and interviews with the creators of many of the most sophisticated systems currently in operation.

Singer divides his work into two key portions, entitled "The Change We Are Creating" and "What Change Is Creating for Us." In the first, he illustrates the history of robotic designs and automated weaponry, with a recurring theme that humans are inherently interested in creating new technologies for the sake of the technologies themselves, with little thought given to the unexpected results of revolutionary innovation. In the second portion, he considers the probable outcomes of incorporating more advanced military technology into the battlefield, and points out that robots being used to fight war are not simply a tactical innovation—their influence is far more pervasive and is felt at every level of the political-military spectrum. To Singer, the decision to move forward is fraught with peril, as the changes that military robots trigger may prove to make warfare infinitely more destructive, more likely, and more unpredictable.

Torrance, S. B., ed. 1984. *The Mind and the Machine: Philosophical Aspects of Artificial Intelligence.* **New York: John Wiley.**

This is a philosophical approach to a highly technical field, the development of artificial intelligence. It is based on papers given at a 1983 conference and thus allows the reader to have a quick snapshot of the field at a relatively early stage of its development. The authors present a number of mental models, and spend a great deal of space differentiating between cognitive and experiential states. Also, they examine some of the theoretical limits of artificial intelligence long before they could be reached, demonstrating that even three decades ago, some of the practical limitations were well understood.

Wagner, William and William P. Sloan. 1992. *Fireflies and Other UAVs.* **Leicester, UK: Midland.**

Wagner served as the historian for the Teledyne Ryan Aeronautical Firm and thus was in the unique position to document the development, and deployment, of the Teledyne Ryan Fire Fly. This book is a thorough examination of the history of the Ryan UAVs, although it does shift topics unpredictably at times and is understandably pro-Ryan. Wagner's access to the Teledyne Ryan test records illustrates some of the challenges facing engineers tasked with creating a radical new technology. The authors present a complete picture of the individuals associated with the creation of a specific platform and the urgent need that was felt to create a reconnaissance aircraft that could photograph the most hostile airspace in North Vietnam.

Warwick, Kevin. 2002. *I, Cyborg.* **London: Century Books.**

Warwick, a relentless self-promoter, envisions a potential world in which machines, smarter and faster than humans, make all of the key decisions that underpin the function of society. However, he also predicts the possibility that humans could enhance themselves, thus maintaining pace with robotic innovations and retaining some measure of control. To promote not only his ideas, but also the science behind them, he utilized himself as a test subject. In 1998, he had a silicon-chip transponder surgically implanted in his arm to demonstrate the human body's capability to integrate neurons with an artificial device. This concept has since been expanded, with experiments to demonstrate the human brain's ability to control computers and even prosthetic limbs.

In 2002, Warwick had an electrode array implanted into the median nerve fibers of his wrist. The electrodes were then used to map specific neurons that controlled muscle movement and sensation in his hand. This allowed him to wirelessly control an external prosthetic arm by moving his own organic limb. He was also able to control an electric wheelchair solely through signals sent from his brain through the electrodes and into a wireless device. When his wife, Irena, had a similar implant placed in her medial nerve bundle, the two were able to send signals back and forth through their electrodes.

Warwick's book describes the series of experiments carried out while the electrodes were implanted and functional, and discusses the possible medical ramifications of his pioneering work. The epilogue of the book discusses Warwick's conception of the future of cybernetics, a deeply disturbing world in which only the best specimens of humanity are chosen for augmentation, and these enhanced humans look down on their "subspecies" brethren who live without implants. In Warwick's vision, unmodified humans of the future are fit only for slave labor, serving the cyborg elites who rule the globe and dominate society.

Journals, Bulletins, and Newsletters

Air & Space Power Journal.

http://www.airpower.au.af.mil/

Air & Space Power Journal is a bimonthly journal produced by the Air University Press, drawing largely on the faculty and students of the Air University schools for strategic analysis of airpower. In line with the U.S. Air Force mission, the publication includes substantial coverage of space and cyberspace issues, as well as articles on the role of new technology in the modern military environment.

Air Force Times.

http://www.airforcetimes.com/

Air Force Times is a weekly newspaper designed for members of the U.S. Air Force but available to anyone who wishes to subscribe. It includes features about new aircraft being added to the Air Force inventory, with many articles dedicated to the increase in unmanned platforms and how they will influence the Air Force mission.

Army Times.

http://www.armytimes.com/

Army Times is a weekly newspaper designed for members of the U.S. Army but available to anyone who wishes to subscribe. It often devotes features to emerging technological issues, including the incorporation of robots and UAVs onto the battlefield, particularly those machines that are being deployed with Army units.

Aviation Week & Space Technology.

http://www.aviationweek.com/

Aviation Week & Space Technology is a weekly magazine produced by McGraw-Hill. It has been in publication since 1947 and has grown to be the largest-circulation aerospace industry publication in the world. The magazine remains primarily focused on aviation, both commercial and military, with a secondary emphasis on the domain of space. It has a large staff, which allows for extremely in-depth reviews of virtually every major issue pertaining to the aerospace field.

Defense News.

http://www.defensenews.com/

Defense News is an independent publication that serves as an aggregating news source for individuals involved in national defense, homeland security, and defense contracting. It devotes substantial coverage to defense politics, both in the United States and abroad, with some coverage of emerging technologies and how they relate to the national strategic situation.

Inside the Air Force.

http://www.airforce-magazine.com/

Inside the Air Force is the magazine of the Air Force Association. It focuses on information of interest to current and former members of the U.S. Air Force and thus has a substantial amount of information regarding new technological innovations that have been adopted by the Air Force, including unmanned platforms.

Navy Times.

http://www.navytimes.com/

Navy Times is a weekly newspaper designed for members of the U.S. Navy but available to anyone who wishes to subscribe. It often covers items of interest to naval personnel serving aboard vessels, including the development of new naval UAVs to assist in the service's mission.

Popular Mechanics.

http://www.popularmechanics.com/

Popular Mechanics is one of the best-known popular magazines devoted to issues of technology. It has been continuously published since 1902 and currently has nine international editions in addition to the original American issue. It often devotes features to the latest technological advancements in the field of robotics, computing, and artificial intelligence, and a section of each edition is dedicated to military technology.

Robot.

http://www.botmag.com/about_robot_magazine.shtml

Robot magazine is designed to stimulate public interest in robotics. It devotes a substantial amount of space to demonstrating projects that can be built in the home but also provides an overview of cutting-edge developments in robotic technology.

Wired.

http://www.wired.com/

Wired is both a print magazine and online periodical, founded in 1993, devoted to emerging technology and its relationship with culture, society, politics, and the economy. It is one of the most widely circulated and well-known technology publications in the world, and has a robust corps of reporters with a reputation for breaking technology stories sooner than their competitors, as well as a large contingent of bloggers who cover virtually every aspect of military technology.

Government Documents and Agency Publications

Baggesen, Arne. 2005. *Design and Operational Aspects of Autonomous Unmanned Combat Aerial Vehicles.* Monterey, CA: Naval Postgraduate School.

Baggesen argues that most future UAV concepts should rely on a swarm concept, with small units being released over a wide area and then moving into autonomous hunter-killer activities. To facilitate this concept, engineers must focus on rapid target detection rates, long attack durations, weapon lethality, target recognition capabilities, and communication. He also believes that because UAVs can be built with lower quality and safety standards, their costs should be much lower than manned systems, possibly to the level that individual UAVs will actually cost less than modern air defense missiles. If so, an enemy's air defense network might be eroded simply by forcing the expenditure of all of their munitions prior to attack by manned aircraft. These small UAVs might also serve in a ballistic missile intercept system, as long as they are capable of detecting, tracking, and intercepting the inbound weapon. The author recognizes that swarm tactics may lead to multiple hits on the same target, or repeated attacks on already-destroyed targets, but he believes an emphasis on minimizing false target attacks, possibly by building in random short attack delays might eliminate simultaneous attacks and keep the swarms cost-effective.

Bone, Elizabeth and Christopher Bolkcom. April 25, 2003. *Unmanned Aerial Vehicles: Background and Issues for Congress.* **Washington, DC: Congressional Research Service. Available at: http://www.fas.org/irp/crs/RL31872.pdf**

This document lays out the history of the UAV program within the Department of Defense, current to 2003. It identifies the major needs, key systems, and massive spending increases that were already underway by that time. In particular, the report is concerned with means to reach the congressional requirement that one third of all deep-strike aircraft should be unmanned by 2010 and offers four potential courses of action to achieve the goal: accelerate the unmanned combat aerial vehicle development program, retire manned aircraft, weaponize current UAVs, or rapidly develop a new combat UAV. The authors note that Congress matched or added every UAV budgetary request in fiscal year 2003, demonstrating broad support for the concept. This work includes an excellent summary of the major UAV platforms in use and should be considered in comparison to the 2012 study by Gertler to see the rapid flux of system acquisitions in the subsequent decade.

Congressional Budget Office. 2011. *Policy Options for Unmanned Aircraft Systems.* **Washington, DC: Author.**

This official report is designed to outline the budget-neutral acquisitions options available to Congress for the procurement of military unmanned aircraft. It compares the plans held by each of the four services and considers whether such plans should be approved, modified, or denied by Congress, based on both system capabilities and fiscal constraints. The fiscal data contained within the document show the relatively low cost of the UAV programs, relative to the size of the overall Department of Defense budget, but also demonstrate the expectations of rising costs for the next decade. Many of the assumptions within the document are based on the possibility of designing a new UAV that would combine the size and range of the RQ-4 Global Hawk with the payload capacity and strike capabilities of the MQ-9 Reaper.

Department of Defense. 2008. *National Defense Strategy.* **Washington, DC: Office of the Secretary of Defense.**

http://www.defense.gov/

The *National Defense Strategy* outlines the role to be played by the Department of Defense and each of the military services in fulfilling the strategic vision outlined by the *National Security Strategy.* It describes how the Department of Defense will carry out the objectives created by the commander in chief, including carrying out current conflicts, preventing future threats, avoiding the proliferation of weapons of mass destruction, and modernizing military institutions to meet the challenges of the future. Previous editions of the document are available through the same website.

Department of Defense. 2010. *Quadrennial Defense Review Report.* **Washington, DC: Office of the Secretary of Defense.**

http://www.defense.gov/qdr/

The *Quadrennial Defense Review,* as the title implies, is completed every four years in accordance with a legislative mandate that calls for the Department of Defense to review and update its strategy and priorities. It helps the Pentagon maintain a long view regarding the major threats and challenges, while also ensuring that procurement, training, and personnel acquisition are well balanced to meet the threats and conflicts that the nation faces. Previous versions of the report are available at the same website.

Drezner, Jeffrey A. and Robert S. Leonard. 2002. *Innovative Development: Global Hawk and Dark Star,* **4 vols. Santa Monica, CA: RAND.**

This RAND study compares two different high-altitude, long-endurance UAV programs that were both initially sponsored by the Defense Advanced Research Projects Agency (DARPA) and the Defense Airborne Reconnaissance Office. Interestingly, despite being the eventual end-user of one or both platforms, the U.S. Air Force showed almost no interest in either development project, which helps to explain why the authors argue that United States has had three decades of poor results with UAVs. By closely comparing the development cycles of the two aircraft, the work demonstrates how even clear program expectations can result in cost overruns, failed technology demonstrations, and cancelled projects. Ultimately, Global Hawk was chosen after more than 100 months of effort, a longer development cycle than the RQ-1 Predator, the F-117 Nighthawk stealth fighter, or the F-16 Falcon. In comparison, Dark Star, which was ready earlier than expected but performed poorly in operational tests, was eventually cancelled by the Air Force after only a handful of test flights, something the authors suggest demonstrates the Air Force did not want either program to go forward but could kill only one of them.

Gertler, Jeremiah. January 3, 2012. *U.S. Unmanned Aerial Systems*. Washington, DC: Congressional Research Service. Available at http://www.fas.org/sgp/crs/natsec/R42136.pdf.

Gertler has written a summary of the current state of UAV systems in the Department of Defense. He notes that there has been massive growth in spending on unmanned aircraft, with the total expenditures rising from $284 million in fiscal year 2000 to $3.3 billion in fiscal year 2010. In that same period, the American inventory of unmanned aircraft increased by 4,000 percent. In addition to a synopsis of the major systems currently in use, Gertler provides some background on potential problems within the fleet, including the exponentially growing cost of sensors, failure of systems to function with one another, and lack of bandwidth. As a government report, this document is unsurprisingly based in fiscal considerations, but it is an excellent source of cost comparison data for the various systems.

Guetlein, Mike. 2005. *Lethal Autonomous Weapons: Ethical and Doctrinal Implications*. Newport, RI: Naval War College.

Guetlein sees many advantages to the idea of granting autonomous lethal decision making to unmanned platforms. Essentially, he believes it will result in faster, better, cheaper military systems, particularly those designated for precision engagements of ground targets. However, he concedes that autonomy might lead to information overload, crowded bandwidths, and a failure to include nonmilitary policymakers in military decisions. He also calls for the United States to assume the lead role in establishing the legal parameters governing autonomous weapons, before they become a common element on the battlefield.

Haulman, Daniel. 2003. *U.S. Unmanned Aerial Vehicles in Combat, 1991–2003*. Maxwell Air Force Base, AL: U.S. Air Force Historical Research Agency.

Haulman offers a series of recommendations for future UAV systems to reduce their loss rates and increase their utility. He argues that UAV flights need to be synchronized with one another and flights of other systems to avoid crashes, and the individual platforms need increased resistance to poor weather as well as mechanical and communication failures. While he finds unmanned aircraft to be extremely useful, they have also proven extremely vulnerable to enemy action, with at least 16 shot down by enemy air defenses in the period from 1995 to 2002. He believes greater specialization in system designs would enable a greater variety of missions to be performed well.

Huber, Arthur F., II. 2002. *Death by a Thousand Cuts: Micro-Air Vehicles in the Service of Air Force Missions*. Occasional Paper No. 29. Maxwell Air Force Base, AL: Center for Strategy and Technology, Air War College.

Huber defines micro-air vehicles as UAVs with wingspans of approximately 15 centimeters or less, but not so small as meso- and nano-air vehicles. He envisions the future of these devices as ISR platforms, using swarming technology and possibly including an electronic disabling system. DARPA is currently funding research in this field, to the great interest of all of the services except the Air Force. The author believes that these types of vehicles will be in active service within a few years of his thesis, which proved true in an experimental sense, but such machines have not seen significant operational usage to date.

Hume, David B. 2007. *Integration of Weaponized Unmanned Aircraft into the Air-to-Ground System.* **Maxwell Paper No. 41. Maxwell Air Force Base, AL: Air War College.**

Hume presents a military territoriality argument that because unmanned aircraft fulfill a fundamental U.S. Air Force (USAF) mission, they should be entirely under USAF control, with joint doctrine for their employment. His primary argument is that crowded airspace and overlapping acquisitions processes might otherwise lead to a greater number of accidents or duplication of procurement efforts. The work also includes an overview of the most prominent systems in use in 2007, and the fundamental strengths and weaknesses of each, from the Air Force perspective. In an interesting concession to the U.S. Army, Hume advocates allowing enlisted personnel to operate UAVs, but only if those personnel are in the Air Force.

Obama, Barack. 2010. *National Security Strategy.* **Washington, DC: Office of the President. Available at http://www.whitehouse.gov/ sites/default/files/rss_viewer/national_security_strategy.pdf.**

The *National Security Strategy* is the overarching guidance provided by the president of the United States to the entire federal government and all of the agencies that contribute to the defense and security of the nation. It outlines the key challenges to national security and the primary means by which the government will address those challenges. It incorporates all of the powers the administration can bring to bear upon the problems, and guides and informs the *National Defense Strategy,* the *National Military Strategy,* and other key guiding documents.

Pescovitz, David. 2010. *Small Unmanned Aerial Systems and the National Air Space: How the U.S. Government Can Spur DIY Drone Innovation.* **Washington, DC: Science and Technology Policy Institute.**

The author envisions a program to spur civilian unmanned aircraft experimentation and development for nonmilitary applications, including disaster response, agriculture, search and rescue, pollution and environmental studies, transport, and traffic monitoring. He identifies the large community of citizens already interested in flying model aircraft and modifying their vehicles to fulfill new roles. Pescovitz notes that many hobbyists are capable of impressive innovation and have created sophisticated designs for extremely low costs. More than anything, he sees the Federal Aviation Administration as the key roadblock to such a program, due to overarching safety concerns that he believes could be fairly easily overcome. He calls for a Grand Challenge for Drones, in much the same manner that the Defense Advanced Research Projects Agency stimulated unmanned ground vehicle research through the establishment of a competition.

Pierce, George M., II. 2000. *Robotics: Military Applications for Special Operations Forces*. Maxwell Air Force Base, AL: Air Command and Staff College.

This author deliberately attempts to publicize the existence of tactical mobile robots, hoping that such publicity will generate both enthusiasm for their development and a discussion of how to properly employ them. These types of robots are primarily designed to precede special operations forces, essentially serving as inorganic scouts sending back imagery of the tactical situation in their immediate surroundings. Pierce lays out a series of necessary attributes for tactical robots, including the ability to restore their own lost communication, the ability to right themselves or operate upside-down, an antihandling self-defense mechanism, and the ability to negotiate stairs, particularly for operations in urban environments.

Sundvall, Timothy J. 2006. *Robocraft: Engineering National Security with Unmanned Aerial Vehicles*. Maxwell Air Force Base, AL: School of Advanced Air and Space Studies.

This thesis argues that combat effectiveness is the primary factor in the decision to field a particular system, not the question of whether it is a manned or unmanned system. The author believes that the line between cruise missiles and UAVs is beginning to blur, due to technological innovation, as is the line between rated pilots and UAV specialists. Sundvall also argues that as machines approach sentience, the need for a pilot will gradually reduce, and the machines will be granted more autonomy to perform their missions without human interference.

Swinson, Mark L. 1997. *Battlefield Robots for Army XXI*. Carlisle Barracks, PA: U.S. Army War College.

Swinson argues that unmanned ground vehicles are the obvious solution to problems in clearing land mine–infested areas.

Despite the potential role, though, he argues that most robotic research has gone into the production of aerial and underwater systems, in part because the U.S. Army has not placed significant emphasis on developing its own robotic platforms. The author also calls for the creation of small, mobile, smart land mines capable of moving to engage specific targets that come within range, despite international calls for the banning of land mine usage.

Trefz, John L., Jr. 2003. *From Persistent ISR to Precision Strikes: The Expanding Role of UAVs*. Newport, RI: Naval War College.

Trefz offers a series of recommendations for how UAVs should be improved in the early 21st century. Specifically, the military needs more satellites to provide enough bandwidth for the control of all of the expanding units, while also developing logarithms that will reduce bandwidth requirements. UAVs need redundant, reliable systems to reduce accident rates. They should continue to perform the dirty and dangerous tasks that manned aircraft should not perform. Commanders need a better understanding of UAV capabilities, which might be provided by incorporating UAV activities into joint warfare publications. Finally, the military needs to continue testing new-concept UAVs, rather than locking into a single platform.

Van Joolen, Vincent J. 2000. *Artificial Intelligence and Robotics on the Battlefields of 2020?* Carlisle, PA: Army War College.

Van Joolen argues that robots utilizing artificial intelligence are innovative but immature technologies in 2000, but that they will soon come to dominate the modern battlefield. In part, he bases his argument on the continually increasing processing speed of computers, which has already exceeded humans and shows no signs of slowing down. Three key assumptions underpin the argument. First, every 20th-century weapon, including nuclear warheads, will be available to enemy forces. Second, dominance in space will lead to dominance in the electromagnetic spectrum. Third, dominating the electromagnetic spectrum will lead to superior battlefield awareness and operational advantages. Thus, victory in 2020 will depend on the possession of new technology, space dominance, and a willingness to utilize both.

Williams, Kevin W. 2004. *A Summary of Unmanned Aircraft Accident/Incident Data: Human Factors Implications*. Washington, DC: Department of Transportation and Federal Aviation Administration.

Williams evaluated the human factors in UAV accidents and found that human error was responsible for up to 68 percent of accidents, depending on the platform. For most systems, electromechanical failure proved the most common trigger for accidents. The accident rate for UAVs is much higher than that for manned aircraft, in part due to a lack of redundant safety mechanisms. The comparisons within the document illustrate that the producers of UAVs are often not primarily aircraft manufacturers, and thus their designs do not follow typical aviation display concepts. The result is often confusion among human operators, which leads to an abysmal accident rate.

Young, Peter W. 1986. *Autonomous Unmanned Vehicles for Air Warfare*. Maxwell Air Force Base, AL: Air Command and Staff College.

Young argues in favor of UAV development for aerial warfare, primarily as a counter to the improving capability of America's Cold War rivals. Given the increase in the Soviet manned aircraft program, both qualitatively and quantitatively, the author believed that unmanned aircraft might be able to maintain the American advantage in aerial warfare. This thesis was produced before UAVs became a common part of the U.S. military inventory and successfully projected a much greater role for them as computing power continued to increase.

Young, Virginia. 1996. *Weaponization Concepts for Unmanned Systems*. Redstone Arsenal, AL: U.S. Army Missile Command.

Young examines the future utility of unmanned ground vehicles in the U.S. Army Modernization Plan. She identifies their primary responsibilities to be reconnaissance, security, explosive ordnance disposal, and mining/countermining operations. To survive budgetary constraints, the robots need to be tested to demonstrate their cost/benefit ratio was superior to existing methods. Given the personnel reductions in place during the 1990s, it is unsurprising that the author pushes for robots as a force multiplier, although her vision of an antiarmor function is fairly rare. In addition, the author argues that remote-controlled High Mobility Multipurpose Wheeled Vehicles (HMMWVs) could be used to transport automated air defense batteries or antiarmor munitions quickly and efficiently.

Web Sites

Air Force Research Laboratory

http://www.wpafb.af.mil/AFRL/

The Air Force Research Laboratory has an online presence that includes unclassified information about ongoing research projects and technological innovations as well as the effects of new technologies in the Air Force inventory. It often serves as a clearinghouse to announce new platforms that have been adopted into service.

Carnegie Mellon University Robotics Institute

http://www.ri.cmu.edu/

The Carnegie Mellon University (CMU) Robotics Institute conducts substantial online outreach, encouraging members of the public to learn more about ongoing academic research into robotic capabilities, and to occasionally participate in robotics projects. The site also offers a substantial amount of history about the field of robotics and the role that CMU has played in its development.

Defense Advanced Research Projects Agency (DARPA)

http://www.darpa.mil/

The DARPA website provides a surprising amount of transparency regarding ongoing projects and areas of interest. Of course, it does not offer access to technical details of each program, but it does illustrate the various concepts being investigated by DARPA researchers, allowing the user to examine the types of revolutionary leaps forward being pursued.

Defense Daily Network

http://www.defensedaily.com/

Defense Daily is a trade publication designed for defense contractors and people in related fields. Their online website is a clearinghouse for virtually every aspect of the political-military relationship, with a particular emphasis on contracting, subcontracting, emerging markets, and international defense decisions.

iRobot Corporation

http://www.irobot.com/

The iRobot Corporation website is a dual-purpose site dedicated to marketing the robots designed and sold by the corporation and to providing information to educators, students, and the general public about the state of the field of robotics. Visitors can purchase any of the home-convenience robots developed by the corporation, and request technical support and service for the products. Visitors can also examine the various military and maritime robots the corporation has developed and deployed around the globe. The website also provides information about the latest research agendas of corporate employees and discusses how the iRobot corporation supports robotics education at all age levels.

YouTube

http://www.youtube.com/

One of the easiest locations to see the most recent robotic innovations is on YouTube, where footage of virtually every form of robot and UAV in any military in the world can be found. Often, the footage is not officially released, but rather shot by observers who are not part of the military using the technology. Nevertheless, it is in the public domain and easily accessed from virtually any Internet terminal. Also, sensor footage taken by UAVs and robots is often posted, allowing the viewer to develop a much better understanding of how robots perceive the world.

Glossary

AI   Artificial Intelligence

Android   A robot designed to look and act like a human being.

ASIMO   Advanced Step in Innovative Mobility

ATE   Advanced Technologies & Engineering; South African corporation specializing in UAVs

ATS   Advanced Target Systems; Primary UAV developer for the United Arab Emirates

Automaton   A moving mechanical device made to imitate a human being, which usually performs a function defined by a specific coded set of instructions.

BWAST   Beijing Wisewell Avionics Science and Technology Company

CASIC   China Aerospace Science and Industry Corporation

CIA   Central Intelligence Agency

CIWS   Close-in Weapons System

CRAM   Counter Rocket Artillery Mortar

CSIST   Chang Shan Institute of Science and Technology; UAV research facility in Taiwan

CTRM   Composites Technology Research Malaysia

Cybernetics   The study of the structure of regulatory systems, including information and systems theory

Cyborg   Cybernetic organism; an individual incorporating both organic and artificial components

DARPA   Defense Advanced Research Projects Agency

DIA   Defense Intelligence Agency

DOD   Department of Defense

EOD   Explosive Ordnance Disposal

FCS   Future Combat System

FEMA   Federal Emergency Management Agency

Firescout   A rotary-winged remotely piloted aerial vehicle

GAIC   Guizhou Aviation Industry Group

GCCS   Global Command and Control System

GPS   Global Positioning System

HALE   High Altitude, Long Endurance

HESA   Iran Aircraft Manufacturing Industries

HMMWV   High Mobility Multipurpose Wheeled Vehicle. Commonly called a Humvee.

IAI   Israel Aerospace Industries

IDF   Israel Defense Forces

IED   Improvised Explosive Device

IREX   International Robotics Exhibition

ISR   Intelligence, Surveillance, and Reconnaissance

IT   Information Technology

JAI   Jordan Aerospace Industries

JARS   Jordan Advanced Remote Systems

KAI   Korea Aerospace Industries

KARI   Korea Aerospace Research Institute

KHZ   Kilohertz. A measure of frequency defined as one thousand cycles per second of a phenomenon.

MARCBOT   Multifunction-Agile Remote-Controlled Robot

MHZ   Megahertz. A measure of frequency defined as one million cycles per second of a phenomenon.

MiG   Mikoyan and Gurevich Design Bureau; Soviet and Russian aircraft design and production company

MILE   Medium Altitude, Long Endurance

Moore's Law   States that the number of electronic components that can be fitted on a single integrated circuit (microchip) doubles each year, exponentially increasing the computing power of electronic devices; similar growth has been identified in other high-technology fields, including robotics

MULE   Multifunction Utility/Logistics and Equipment Vehicle

NATO   North Atlantic Treaty Organization; a defensive alliance of 28 states in Europe and North America, first founded in 1948 and most recently expanded in 2009

NDC National Development Complex (NDC); Pakistani corporation specializing in UAVs Neural Network

NRIST Nanjing Research Institute on Simulation Technique

NSC National Security Council

NSF National Science Foundation

OODA Observe, Orient, Decide, and Act

OSD Office of the Secretary of Defense

QAI Qods Aviation Industries; Iranian UAV developer

RAC MiG Russian Aircraft Corporation MiG; *see also* "MiG"

RPV Remotely Piloted Vehicle

RSA Republic of South Africa

SEAL Sea, Air, and Land. Acronym for U.S. Navy special forces.

SWORDS Special Weapons Observation Reconnaissance Detection System

TAI Turkish Aerospace Industries

TUGV Tactical Unmanned Ground Vehicle

Turing Test Test of artificial intelligence first proposed by Alan Turing in 1950

UAE United Arab Emirates

UAV Unmanned Aerial Vehicle

UCAV Unmanned Combat Aerial Vehicle

UGV Unmanned Ground Vehicle

USA United States Army

USAF United States Air Force

USB Universal Serial Bus, a type of connector between computers and electronic devices.

USMC United States Marine Corps

USN United States Navy

UUV Unmanned Undersea Vehicle

WWAST Wounded Warrior Amputee Softball Team

Index

About the Author

Paul J. Springer is a professor of comparative military studies at the Air Command and Staff College, where he teaches courses on leadership, strategy, terrorism, and military history. He previously taught at the U.S. Military Academy at West Point and Texas A&M University, where he earned his doctorate in military history in 2006. He is the author of *America's Captives: Treatment of Prisoners of War from the Revolution to the War on Terror* (University Press of Kansas, 2010).